Models of
Oculomotor Control

Models of
Oculomotor Control

George K. Hung
Rutgers University, USA

World Scientific
New Jersey • London • Singapore • Hong Kong

Published by

World Scientific Publishing Co. Pte. Ltd.
P O Box 128, Farrer Road, Singapore 912805
USA office: Suite 1B, 1060 Main Street, River Edge, NJ 07661
UK office: 57 Shelton Street, Covent Garden, London WC2H 9HE

British Library Cataloguing-in-Publication Data
A catalogue record for this book is available from the British Library.

MODELS OF OCULOMOTOR CONTROL

Copyright © 2001 by World Scientific Publishing Co. Pte. Ltd.

All rights reserved. This book, or parts thereof, may not be reproduced in any form or by any means, electronic or mechanical, including photocopying, recording or any information storage and retrieval system now known or to be invented, without written permission from the Publisher.

For photocopying of material in this volume, please pay a copying fee through the Copyright Clearance Center, Inc., 222 Rosewood Drive, Danvers, MA 01923, USA. In this case permission to photocopy is not required from the publisher.

ISBN 981-02-4568-8

Printed in Singapore by Uto-Print

To my wife Xiaosi Yang
for her love, devotion, and support

Preface

Biomedical engineers often have difficulty understanding visual systems concepts, whereas non-engineers are often overwhelmed by the apparent complexity of biomedical applications of control theory. There is, moreover, lack of a book or reference source with a balanced treatment of these two needs. It is with this in mind that I have put together a coherent and relatively comprehensive monograph on oculomotor control, based primarily on my research work over the past 20 years. This monograph has two main objectives. First, it aims to provide the biomedical engineer with a thorough understanding of how various engineering control principles are applied to oculomotor systems. Second, it aims to provide the non-engineer with the fundamentals of control theory, and then leads them to understand how various physiological and clinical concepts can be represented quantitatively and efficiently by control systems models.

Oculomotor research involves a wide variety of disciplines. For example, it includes biomedical engineering, and the related fields of biophysics and mathematics. Non-engineering disciplines include neurology, ophthalmology, optometry and psychology. It is because of this breadth of disciplines that a unified and systematic approach is needed for a clearer understanding of oculomotor control. I believe this has been accomplished in the monograph. First, a glossary and introductory chapters on anatomy and physiology of eye movements and basic control systems concepts provide the necessary background. Then, the monograph applies these concepts to static linear and nonlinear analysis of various oculomotor systems. In addition, it presents advanced topics on the application of dynamic linear and nonlinear modeling techniques to the oculomotor system, with a particular emphasis on myopia development, which is an important international health concern.

This monograph has been enriched by the clinical and logical insights brought forth by Dr. Kenneth J. Ciuffreda, who has collaborated on a number of articles cited in the monograph. In addition, Dr. Bai-Chuan Jiang has contributed substantially to the discussions on interactive accommodation and vergence system. It is hoped that the monograph will become a valuable reference for both bioengineers and vision scientists, and serve as the foundation for continued research in the fascinating and complex field of oculomotor control.

George K. Hung
Department of Biomedical Engineering
Rutgers University

Acknowledgements

The author wishes to thank Dr. Kenneth J. Ciuffreda, State College of Optometry, State University of New York, New York City, for his contributions to some of the articles discussed in this monograph and his valuable comments, and to Dr. Joseph Wilder, Computer Aids for Industrial Productivity, Rutgers University, Piscataway, New Jersey, for his valuable comments. The research has been supported over the years by grants from Essilor International and NIH–EY03709, 07519 and 08817.

Contents

Preface vii

Acknowledgements ix

Introduction 1
    Basic Anatomy and Physiology of Eye Movements 5
    Basic Measurement Terms 7
    Basic Control System Concepts 10
    Eye Movement Measurement Techniques 15
        Stimulus arrangement and typical experimental protocol 15
        Accommodation measurement 15
            Static 15
            Dynamic 17
        Dynamic vergence and saccadic eye movement measurement 18

Static Analysis Techniques 23
    Accommodation System 23
    Vergence System 35
    Linear Analysis of Relationship Between AC and ACG 35
    Nonlinear Analysis of AC/A Using the Phoria and Fixation
      Disparity Methods 39
        Derivation of model equations 41
            Open-loop vergence 41
            Closed-loop vergence 43
        Model simulations 46
    Proximal Model 49
    Sensitivity Analysis of Accommodation and
      Vergence Interactions 56

Dynamic Analysis Techniques — 61

Main Sequence — 61
Accommodation System — Root Locus Analysis — 61
Vergence Dual-Mode Dynamic Model — 63
Accommodative Dual-Mode Dynamic Characteristics — 73
Adaptation Model of Accommodation and Vergence — 76
Nearwork-Induced Transient Myopia (NITM) Model — 81
Refractive Error Development Model — 87
   Background — 87
   Incremental retinal-defocus theory — 89
      Corneal growth does not contribute to the emmetropization process after two years of age — 90
      Neuromodulators control sensitivity to changes in retinal-image contrast — 90
      The overall mechanism for regulating the rate of axial length growth — 91
   Applications of the theory — 92
      Lenses — 92
      Prolonged nearwork — 93
   Basic retinal anatomy and physiology — 96
   Model of refractive error development — 96
Saccade-Vergence Interactions Dynamic Model — 102
Summary Remarks — 110

References — 113

Index — 125

Introduction

Eye movements are accurate reflections of the brain's control strategy. Their function is to provide essential information about the visual scene under a wide variety of situations that are encountered in our daily lives. Therefore, understanding how eye movements are controlled in both normal and symptomatic individuals is one of the most important goals of vision scientists and bioengineers. This monograph provides a summary of significant research results on the quantitation of oculomotor behavior using control systems analysis techniques, based primarily on my research work over the past 20 years.

When one redirects gaze, neural command signals change the lens focus and rotate the two eyes to provide a clear and single image of the new target. It has been found that three primary oculomotor movements are involved in the automatic control of binocular gaze. *Accommodation*, or focusing, changes the lens power in response to change in depth of the target; *vergence* rotates the eyes symmetrically in opposite directions to point accurately at the binocularly fixated target; and *saccades* rotate the eyes in the same direction to accurately adjust for the lateral displacement of the target. These automatic adjustments can be represented in engineering block diagram form as three feedback control systems. Well-known engineering control systems theories can be used to study the feedback control of the models of these physiological processes.

This monograph is intended for both the engineer with relatively little physiology background and the vision scientist with relatively little engineering background. Basic physiology of eye movements as well as basic control systems theory are introduced to provide the background needed by the reader to understand the application of various engineering techniques to eye movement control systems. Due to the numerous control systems and visual science terms used, a table containing a glossary of terms has been included (Table 1). The monograph is divided into three sections. The "Introduction" reviews basic anatomy and physiology of eye movements, basic measurement terms, basic control systems concepts, and accommodation and eye movement instrumentation and measurement methodology. The "Static Analysis Techniques" section examines in detail simulations of the static

2 Oculomotor Control Models

Table 1 Glossary of Terms
(Adapted from Hung,[65] p. 306.)

Terms	Definition
ABIAS	Accommodative bias or tonic level.
AC	Accommodative convergence crosslink gain.
AC/A	Accommodative convergence to accommodation ratio.
Accommodation	A change in the optical power of the lens to minimize the retinal defocus.
ACG	Accommodative controller gain.
AD	The deadspace limit value for depth of focus.
AE	Accommodative error.
Amblyopia	Amblyopia is a reduction in monocular visual acuity that is not correctable by refractive means and is not attributable to obvious structural or pathological ocular anomalies.
Anomalous retinal correspondence	A type of correspondence between the two retinas, occurring frequently in strabismus, in which the fovea of one eye does not correspond to the fovea in the other eye, but instead to an extrafoveal area in that eye.
AR	Accommodative response.
AS	Accommodative stimulus.
BI	Base-in prism, providing a divergent stimulus.
$B_{nonlinear}$	Intercept term representing 16 possible combinations.
BO	Base-out prism, providing a convergent stimulus.
C	Subscript denoting a constant viewing distance.
CA	Convergence accommodation crosslink gain.
CA/C	Convergence accommodation to convergence ratio.
Deadspace operator	An ideal mathematical function that approximates the threshold of detection of a variable in a physiological system. If the absolute value of the input to the function is less than the threshold (with threshold being a positive number), then the output is zero. On the other hand, if the input is greater than the threshold, then the output is equal to: input − threshold. Further, if the input is less than −threshold, then the output is equal to: input + threshold.
Diopter (D)	A unit of optical power equal to the reciprocal of the distance of the target from the corneal plane of the subject measured in meters.

Table 1   (Continued)

Term	Definition
Depth of focus (DOF)	The deadspace operator for accommodation. It is the range of an image, either in front of or behind the retina, that is perceived to be clear and sharp. It is assumed that this range corresponds to the limits in the size of the blur circle on the retina which can be tolerated without the perception of appreciable amount of blur. The DOF varies with pupil size, ranging from $\pm\,0.15$, $\pm\,0.30$ to $\pm\,0.80$ D for 8, 3 and 1 mm diameter pupil, respectively.
Emmetropia	Normal refractive condition in which distant objects are focused on the retina when accommodation in minimally stimulated.
Emmetropization	A change in the rate of axial growth which compensates for and reduces the effect of imposed retinal defocus, usually over a relatively long time interval.
Fixation disparity (FD)	Equal to the vergence error (VE), or difference between vergence stimulus (VS) and response (VR), under the binocular viewing condition. Eso FD describes a response which is more convergent than the stimulus, and Exo FD describes a response which is less convergent than the stimulus.
Hyperopia	A refractive condition in which distant objects are focused behind the retina when accommodation in minimally stimulated.
Hysteresis	A difference in the response path for stimulus in the forward and reverse directions, such as that seen in the magnetization curves.
L	Lens value; or subscript denoting the lens viewing condition.
Meter angle (MA)	A measure of vergence angle equal to the reciprocal of the distance of a target from the centers of rotations of the eyes of a binocularly viewing subject measured in meters. The interpupillary distance is implied in the use of MA.
Myopia	A refractive condition in which distant objects are focused in front of the retina when accommodation in minimally stimulated.
Nystagmus	A regular, repetitive, and usually rapid involuntary movement or rotation of the eye that is either oscillatory or has slow and fast phases in alternate directions.
P	Prism value; or subscript denoting the prism viewing condition.

Table 1 (Continued)

Term	Definition
Panum's fusional area (PFA)	The deadspace operator in vergence. It is an area on the retina in one eye, which when stimulated simultaneously with a single specific point on the retina in the other eye, will give rise to a single fused percept. Its diameter increases with target eccentricity, and ranges from about 6–12 min of arc near the fovea to about 20–40 min of arc at 15 deg eccentricity.
Phoria	The angle of the two eyes in the absence of adequate fusion relative to binocular fixation on the target (e.g. one eye blocked). Esophoria describes an angle of the two eyes which is more convergent than exact alignment of the two eyes with the target. On the other hand, exophoria describes a more divergent angle than exact alignment with the target.
Saccade	Rapid same-direction rotations of the two eyes to fix on a target that is displaced laterally in space, such as during reading.
Strabismus (or tropia)	An anomaly of binocular vision under normal viewing conditions wherein one visual axis fails to intersect the object of regard, and thus bifoveal fixation is not attained (cross-eyed).
VBIAS	Vergence bias or tonic level.
VCG	Vergence controller gain.
VD	The deadspace limit value for Panum's fusional area.
VE	Vergence error.
Vergence	Oppositely-directed rotations of the eyes to bring them into alignment with a target in depth.
VR	Vergence response.
VS	Vergence stimulus.

accommodation and vergence systems, linear and nonlinear analysis of crosslink interactions, contributions due to proximal effects, and sensitivity analysis of accommodation and vergence to parameter variations. The "Dynamic Analysis Techniques" section examines in detail the main sequence as a tool for characterizing the dynamic characteristics of the three oculomotor systems, accommodative root locus stability analysis, vergence dual-mode model, accommodation dual-mode characteristics, adaptation model of accommodation and vergence, nearwork-induced transient myopia (NITM) model, refractive error development model and emmetropization, and model of saccade-vergence interactions.

Basic Anatomy and Physiology of Eye Movements

The goal of the accommodation, or focusing, system is to provide a clear and sharp image of an object on the retina. Figure 1A shows a cross-sectional view of the interior of the human eyeball. Light rays enter the eye first through the transparent cornea, which comprises about 2/3 of the fixed refractive power of the eye. The rays then pass through the opening in the iris, called the pupil, and is refracted by the transparent lens, which comprises the remaining 1/3 of the fixed optical power. The lens has, in addition, a variable component that is controlled by the ciliary muscle (which is part of the ciliary body) through its action via the zonular fibers located between the ciliary body and the lens. In this way, the light rays of a target at different distances can be focused by the variable-powered lens onto the fovea, which is a small high-acuity region on the retina. The act of focusing from a far (F) to a near (N) target is called accommodation (Fig. 1B). Moreover, the image can be focused within a certain range either in front of or behind the retina, thus providing a small amount of retinal-defocus, and still be perceived as clear and sharp. This is called the depth-of-focus or DOF (not explicitly shown in Fig. 1B). However, if the retinal-defocus is outside the DOF, the image is perceived to be blurred, and accommodative feedback is used to change the lens power and reduce this blur to a minimum. In general, for viewing of objects closer than about 1 m, the resultant image is focused behind the retina and the accommodative response is said to "lag" the stimulus. Although retinal-defocus, and hence the perceived blur, is an even-error signal (for stimuli given in diopters rather than linear displacement units)[4] that does not provide a direction sense of the error, other optical cues such as chromatic aberration, where light rays of shorter wavelength (e.g. blue) are refracted more than those of longer wavelength (e.g. red), and spherical aberration, where peripheral rays impinging on a lens are refracted more than central rays, as well as

6   Oculomotor Control Models

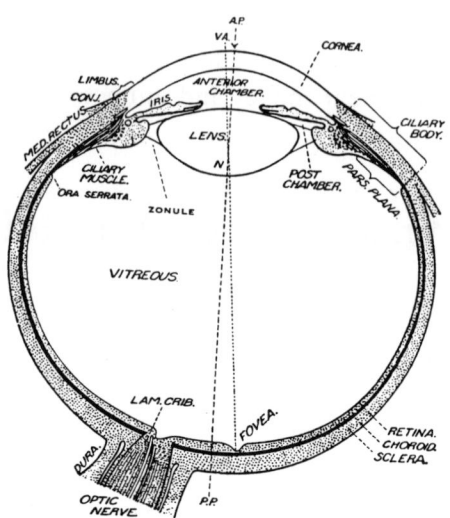

Fig. 1A Cut-out section of the interior of the eyeball showing the transparent cornea and lens, which provide 2/3 and 1/3, respectively, of the refractive power of the eye. The ciliary muscle controls the front surface curvature of the lens to provide variable focusing. The light-sensitive retina contains a small region, called the fovea, for acute vision (adapted from Last,[97] p. 30, with permission).

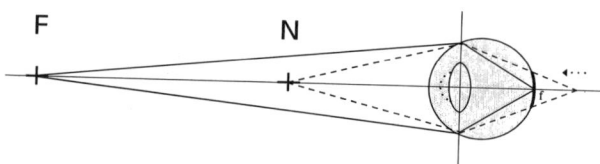

Fig. 1B Schematic diagram of light rays for initially far target (F; solid lines) that is focused on the fovea (f) and the sudden introduction of a near target (N; dashed lines). Accommodation to the near target changes the curvature of the front of lens and moves the under-converged light rays forward to bring the rays into focus at f (see dashed arrow).

perceptual cues (such as size, perspective and overlap) all contribute to accurate direction sense of accommodation in normal daily life. Thus, the initiation of the accommodative response is almost invariably in the correct direction under real-life conditions.

The goal of the binocular eye movement system is to provide a single (rather than double) percept by bringing the images of a target onto corresponding retinal points in the two eyes. Hence, when the target moves in depth, the eyeball in each eye must be rotated by the muscles outside the eyeball, called extraocular muscles (Fig. 2A), to once again bring the images in register on the retinas. There are three pairs of extraocular muscles that are concerned with horizontal, vertical and oblique rotations of the eye. The very efficient pulley actions of these extraocular muscles and their multidimensional control of eye rotations can be fully appreciated by the mechanical engineer. In this monograph, we will be concerned primarily with the horizontal muscles, called the medial rectus and lateral rectus, that are reciprocally innervated and rotate the eye in the horizontal plane. The neural pathways for the control of horizontal eye movements are shown in Fig. 2B. The neural signals are formed in the higher neural centers and then sent to the oculomotor (3rd nerve) and abducens (6th nerve) nuclei, which in turn send signals to the horizontal recti muscles. These signals drive the two eyes in a coordinated fashion so that the lines of sight intersect at the target. The resulting images in the two retinas are combined by the brain to form a single percept. When a target is displaced in depth (e.g. between far (F) and near (N) positions in Fig. 3A), an angular difference between the near and far targets, $\alpha - \beta$ (called disparity), is created and causes the muscles to rotate the two eyes in opposite directions to track it in a disjunctive manner. A disjunctive or vergence response for a target displacement from far to near is called convergence, and that from near to far is called divergence. On the other hand, when a target is moved laterally from side to side (e.g. between positions T1 and T2 in Fig. 3B), the two eyes rotate in the same direction to track it in a conjunctive or conjugate manner. There are two types of conjugate eye movements — saccades that jump to follow rapid target displacements and pursuit eye movements that smoothly follow relatively slowly moving targets.

Basic Measurement Terms

The basic unit of measurement for the focusing or accommodation system is the diopter (D). A diopter is a unit of optical power that is equal to the reciprocal of the distance of the target from the corneal plane (or more

8   Oculomotor Control Models

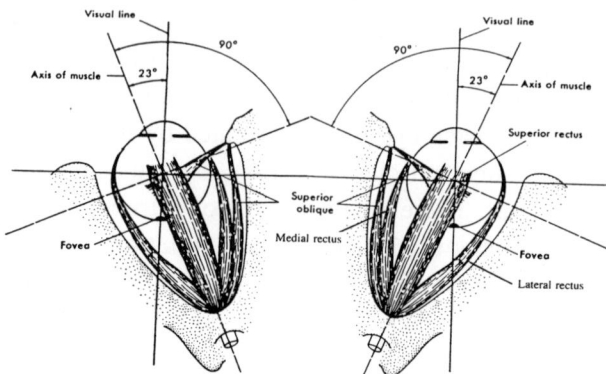

Fig. 2A   The two eyes and the extraocular muscles as seen from above. The orientations of the muscles indicate their lines of action. Horizontal eye rotations, frequently used in daily life, are controlled by the medial and lateral recti muscles. The objective of the muscle-driven eye movements is to bring the target image onto the fovea (adapted from Moses,[111] p. 92, with permission).

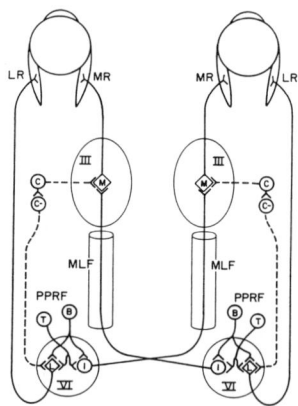

Fig. 2B   Schematic representation of the neural circuitry for the control of horizontal eye movements. Horizontal burst (B) and tonic (T) neurons in the paramedian pontine reticular formation (PPRF) have the signals required for all conjugate horizontal eye movements. These PPRF neurons probably provide similar inputs to both lateral rectus motoneurons (L) and internuclear neurons (I) in the abducens nucleus (VI). Abducens internuclear neurons cross the midline and ascend in the medial longitudinal fasciculus (MLF) to the medial rectus motoneurons (M). A presumed convergence (C) input to the medial rectus motoneurons and its complementary (C-) input to the lateral rectus motoneuron are shown by dashed lines (reprinted from Mays,[103] p. 656, with permission).

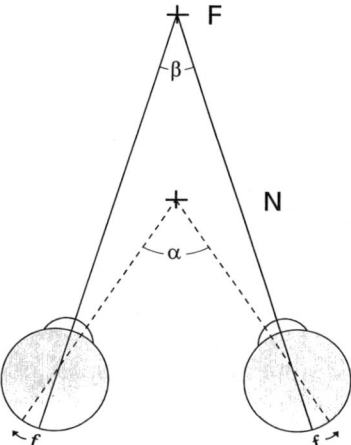

Fig. 3A Binocular fixation of the target at far (F). Rotation of the two eyes in opposite directions (called vergence) brings the near (N) target image onto the foveas. The brain combines the two retinal images into a single percept. The difference in angle between the N and F targets, or $\alpha - \beta$, is called the disparity.

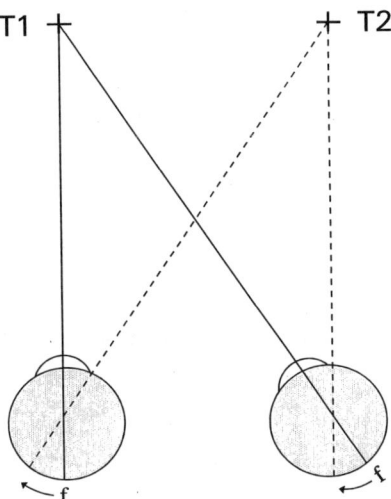

Fig. 3B Binocular fixation initially at target T1. Rotation of the two eyes in the same direction (called version: consisting of fast saccades and slower pursuit eye movements) brings the image of target T2 onto the foveas. Note that the relative angle between the two eyes, or disparity, remains the same.

precisely, the principle plane, which is located 1.35 mm behind the corneal surface[46]) of the subject measured in meters. For example, a target 2 m away requires only 0.5 D of accommodative change, whereas a target 0.5 m away would require 2 D of accommodative change to focus on it clearly.

There are several units of measurement for vergence eye movements. Meter angle (MA), which is used in basic research, is a measure of vergence angle equal to the reciprocal of the distance of a target from the centers of rotations (about 13.5 mm behind the corneal surface[1]) of the two eyes of a binocularly viewing subject measured in meters, and thus is analogous to the diopter used for accommodation. Prism diopter (Δ), which is used in the clinic, is a unit of measure of convergence angle, where 1 Δ is equivalent to 1 cm of lateral displacement at 1 m distance, and is based on the interpupillary distance (PD). Both MA and Δ can be converted to degrees of visual angle.[64] Thus for example, a target 0.5 m in front of a subject with 6.0 cm PD has visual angles of 2 MA (= 1/0.5 m), 12 Δ (= 6 cm * 2 MA), or 6.84 deg (= 2 MA * 6 Δ/MA * 0.57 deg/Δ).

Degrees of visual angle is also used for versional eye movement measurements. The visual angle is the angle formed by the lines of sight from an eye (e.g. left eye) to the two targets (T1 and T2; Fig. 3B).

Basic Control System Concepts

A basic feedback control system block diagram is shown in Fig. 4A. The Laplace operator s is a complex variable equal to $\sigma + j\omega$, which provides information about damping and oscillatory characteristics of a system. The reason for operating in the Laplace domain is that many complicated dynamic operations in the time domain become much simpler mathematical operations

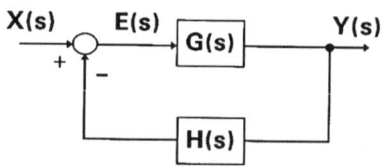

Fig. 4A Block diagram of a simple feedback control system.

in the Laplace domain. For example, when a signal is input to a block, the time domain operation consists of a convolution (i.e. a mathematical operation involving integration of time-shifted functions) between the input and the impulse response of the block. On the other hand, when the signal and the block are transformed to the Laplace domain, the transformed signal and the transfer function of the block are simply multiplied together to provide the output. To simplify this process, one can use transform pairs available in Laplace transform tables as well as a number of stability theories that have been developed. It turns out that it is easier to analyze and interpret a control system's dynamic characteristics in the Laplace than the time domain. When the analysis of the system's characteristics has been completed, the final response can be obtained simply by inverse transformation.

In Fig. 4A, $E(s)$ or the difference between the input $X(s)$ and the product of the output and feedback gain, $Y(s) * H(s)$, serve as the driving signal for the forward-loop gain $G(s)$. It can be shown that the overall transfer function is given by

$$F(s) = \frac{Y(s)}{X(s)} = \frac{G(s)}{1 + G(s) * H(s)}. \tag{1A}$$

For $H(s) = 1$, which is usually the case, we get

$$F(s) = \frac{G(s)}{1 + G(s)}. \tag{1B}$$

It turns out that this apparently simple equation (Eq. 1A) is the basis for much of control systems theory. It can be seen that if $G(s)H(s)$ equals -1, $F(s)$ would go to infinity and the system would become unstable. This can occur if the gain and latency values embedded in $G(s)H(s)$ are too large. Indeed, much of control systems theory involves the determination of the conditions for instability and the system modifications needed to avoid arriving at these unstable conditions.

Another useful control system property is the final value theorem, where the steady-state time domain response is given by:

$$y(t \to \infty) = \lim_{s \to 0} s\, F(s) X(s). \tag{1C}$$

The theorem states that the steady-state time domain step response can be obtained directly from the transfer function. For example, for a unit step input $X(s) = 1/s$, and transfer function $F(s) = 2/(s + 3)$, the steady-state time domain output value is

$$y(t \to \infty) = \lim_{s \to 0} s \, \frac{2}{s+3} \frac{1}{s} = \frac{2}{3}. \tag{1D}$$

The final value theorem forms the basis for *static* analysis of feedback control systems.

A Laplace transform pair that is often used is

$$\frac{1}{s+a} \Leftrightarrow e^{-at} u(t) \tag{1E}$$

where $\Leftrightarrow$ designates transformation between the Laplace and time domains, $u(t)$ is the unit step function, and $a = 1/\tau$, with τ being the time constant (or the time required to reach 63% of the final step response value).

To demonstrate the significant effects of feedback, gain and latency on the dynamics of a system, six MATLAB/SIMULINK block diagrams are shown in Fig. 4B. Starting from the top: (1) a system with a latency element $e^{-T_d s}$, where the latency $T_d = 0.5$ sec, a gain element with gain = 1 and first-order transfer function of the form $1/(\tau s + 1)$, where the time constant $\tau = 0.5$ sec; (2) effect of adding a feedback connection; (3) effect of increasing gain from 1 to 1.5; (4) effect of a further increase in gain to 2.5; (5) effect of increasing latency to 1 sec; and (6) effect of further increase in latency to 2 sec. The simulation outputs (Y1–Y6) are shown in Fig. 4C starting from the upper left figure: (Y1–Y2) shows the significant change in dynamics and steady-state gain when feedback is added; (Y3–Y4) shows larger underdamped response and oscillations as gain is increased; and (Y5–Y6) shows larger underdamped response and oscillations as latency is increased.

In the accommodation and vergence eye movement systems, the latency is long and steady-state gain is high relative to the dynamic response of the system. Hence, if a simple feedback loop was used in a model of the system, the responses would consist of instability oscillations. It turns out that each of the systems solves this problem by separating its control into two parts — a fast open-loop component and a slow closed-loop component. The fast component responds to the stimulus amplitude without feedback to arrive near the desired position. The absence of any feedback (even though sensory input continues to be available) ensures stability of the initial response. Then, a slow closed-loop component takes over and reduces the small residual error to a minimum. Because the residual error is small, the gain of the slow component can be relatively low and still achieve adequate dynamic response. Yet, once these two processes are completed, the overall error would be small and the effective steady-state gain would be equivalent to a continuous feedback system with a high forward-loop gain (with its inherent instability problems). Thus, this dual-mode process achieves both rapid dynamics and small residual error without sacrificing stability.

The saccadic system also has a long latency as compared to its very fast dynamics, and thus would similarly have instability problems if it were in

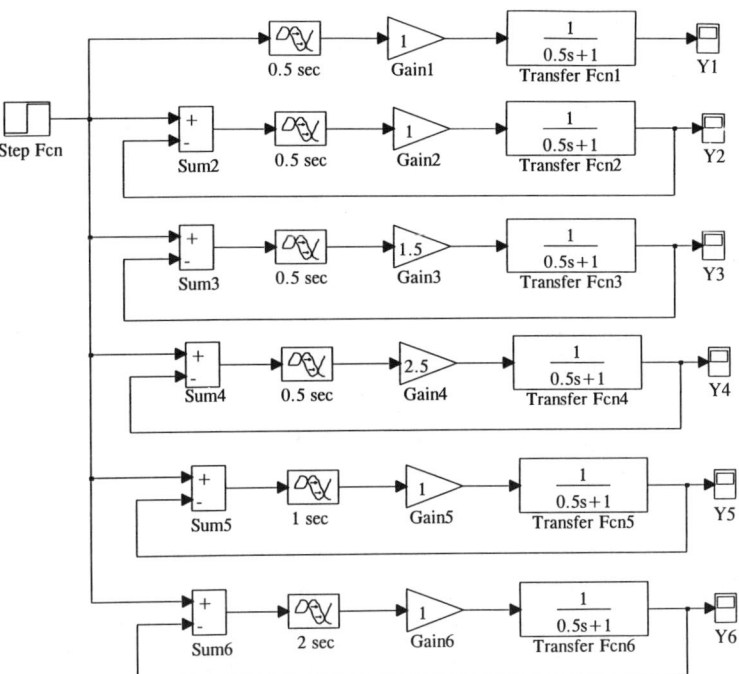

Fig. 4B Six examples of MATLAB/SIMULINK block diagrams demonstrating the effects of feedback, gain, and latency.

14 Oculomotor Control Models

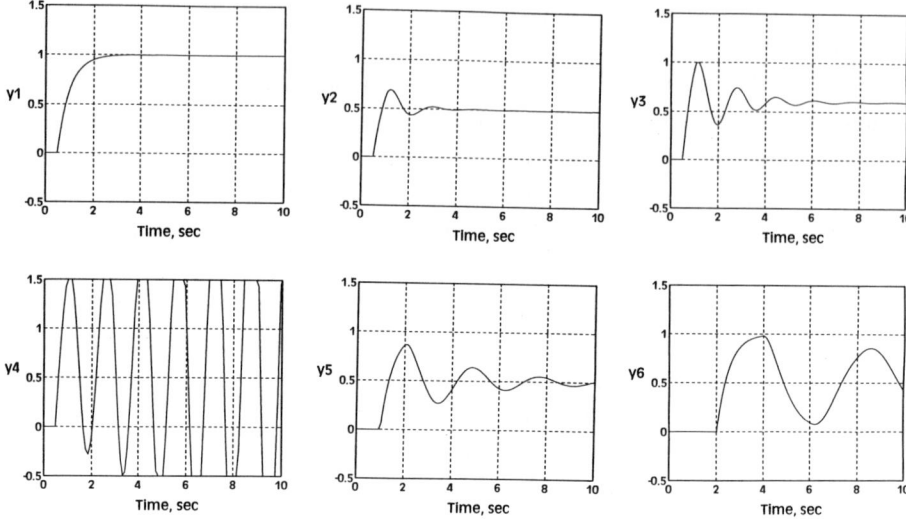

Fig. 4C Simulation results showing significant change due to feedback and increased oscillations in the response as gain and latency are increased.

a simple feedback loop. It solves this problem by responding with an initial open-loop movement, but unlike the accommodation and vergence systems, uses subsequent saccades to make corrective movements.

One of the questions often asked regarding mechanical and physiological feedback systems is "Why can't the response be exactly equal to the stimulus?" There are two ways to answer this question. In terms of a control systems explanation, a stable feedback system will exhibit a small residual error. This can be seen upon examination of Eq. 1B, where even for a very large steady-state gain value for $G(s)$, the overall gain is still slightly less than 1. Thus, the system response cannot be exactly equal to the stimulus. In terms of a physiological explanation, it can be shown that all physical and physiological systems have a threshold for detecting a stimulus change. The region between the upper and lower thresholds is called the deadspace region. A response that falls in the middle of the deadspace region must move within the region until it finally exceeds the threshold and the signal is detected. But if the response dithers just at the boundary of a threshold, it will be able to immediately detect any stimulus change while still maintaining response accuracy. It turns out that the best solution is for the response to be at the boundary of that threshold which requires the least effort for detection, and this occurs when the response is slightly less than the stimulus. Thus, the steady-state response is seen to "lag" the stimulus, rather than being exactly equal to the stimulus. Examples of deadspace in the oculomotor system are the DOF[9] for accommodation and Panum's fusional area (PFA)[121] for vergence.

Eye Movement Measurement Techniques

Stimulus arrangement and typical experimental protocol

Both free- and instrument-space experimental protocols have been used by experimenters.[76] The free-space environment represents the normal visual scene where all the cues such as blur, disparity, overlap, perspective and shading are available to the observer. Moreover, the accommodative and vergence stimuli, based on the reciprocal of the distance of the target from the corneal plane of the observer, are equal or congruent. On the other hand, in the instrument-space environment, many of the cues are purposely removed, so that only blur and/or disparity cues are available to the observer. Moreover, the accommodative and vergence stimuli can be dissociated to provide non-congruent stimuli and a means to "dry dissect"[158] the oculomotor systems.

Accommodation measurement

Static

The static accommodative response can be measured using a Hartinger coincidence optometer. Its operation is based on the Scheiner principle.[54] Consider the situation where the subject's eye is focused at point F but also simultaneously sees the instrument alignment target A, consisting of three long vertical line segments, which is at a position indicated by the cross (Fig. 5). A prism is placed in front of the upper half, but not the lower half, of alignment target A. Thus, the rays emanating from the upper portion of target A (dotted lines) are refracted by the prism and pass through pinhole P1, whereas the rays emanating from the lower half of target A (dashed lines) bypass the prism and traverse directly through pinhole P2. The pinholes result in a large DOF, so that the subject sees the alignment target A as sharp, and does not accommodate for it. Also, since target A is closer to the subject than the focused point F, the rays from A that pass through pinholes P1 and P2 will form two horizontally displaced images on the retina. Due to the prism placed in front of the upper half of target A, the subject sees two sets of short vertical line segments (one upper and one lower) which are displaced horizontally (see bottom illustration in Fig. 5). This technique of using pinholes to present two images of a target on the retina is called the Scheiner principle. As target A is moved closer to F (dashed arrow), the separation between the two images (and therefore the two sets of vertical line segments) is reduced. When A is at F, the two images will be vertically aligned. And, as A is moved beyond F, the two images will once again increase in separation but in the opposite direction. In the Hardinger coincidence optometer, the subject adjusts a dial that moves the alignment target until the images are aligned. The

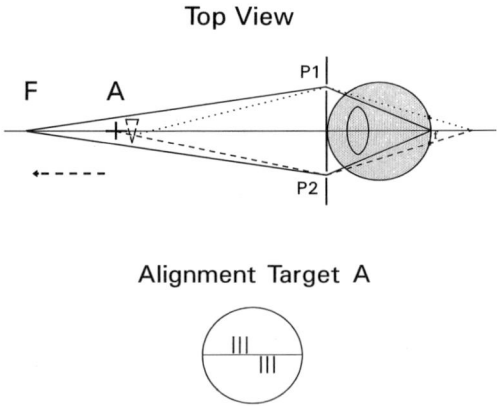

Fig. 5 Scheiner principle and Hardinger optometer. Schematic top-view drawing showing subject focused at F and viewing alignment target A. A prism placed at the upper half (out of page) of target A results in two separate light beams that are projected through two pinholes. Two images are formed on the retina that are displaced horizontally, as shown in the alignment target below. When A is moved to coincide with F, the two images will be aligned vertically. A dial reading corresponding to position of A provides a measure of accommodative response.

dial position can be calibrated to provide a reading of the static accommodative response.

Dynamic

The dynamic accommodative response can be measured using the infrared optometer.[10,29] Its principle of operation is based on the clinical retinoscope, which assesses refractive power of the eye. Figure 6 shows schematic diagrams of the side-view of the eye for the (a) under-, (b) equal-, and (c) over-accommodated conditions. In each diagram, a pinhole aperture in front of target T is swept sequentially in time from position 1, through 2, to 3. The angles of the rays in the diagrams are exaggerated for visual effect, but are meant to represent influence due to lens refraction. If the lens is under-accommodated (A), the target image on the retina will sweep through 1', 2' and 3' from *top to bottom*. However, if the lens is correctly accommodated (B), the target image will remain in one position on the retina. Moreover, if the lens is over-accommodated (C), the target will sweep through 1', 2' and 3' from *bottom to top*. In the dynamic optometer, instead of sweeping across the pinholes, the source is switch back and forth rapidly (150 Hz) between positions 1 and 3. Receiving photodetectors (not shown) that are optically conjugate to the retina will pick up the signals reflected from "top" and "bottom" of the retina. The phase difference of the detector signals provides the direction sense for driving a servomotor, which controls the optical power of the infrared light source, until the phase is zero. The amount and direction of displacement of the servomotor provides a quantitative measure of optical change, or accommodation, which is output as a voltage signal. The resolution of the optometer is typically 0.05 D and the bandwidth is 5 Hz.

Another device, called the SRW-5000 autorefractometer (Shin-Nippon, Japan), uses the grating focus principle. Optics within the device are moved rapidly back and forth by means of electronics to seek the highest contrast in the image of the retinal reflection of the target. This is done in three meridians (0, 45 and 90 deg) so that both spherical and cylinder (associated with astigmatism) corrections are available. The range of the autorefractometer is ±15 D for the spherical reading and ±7 D for the cylinder reading; the resolution is 0.12 D and the reading is provided automatically every 2 sec.

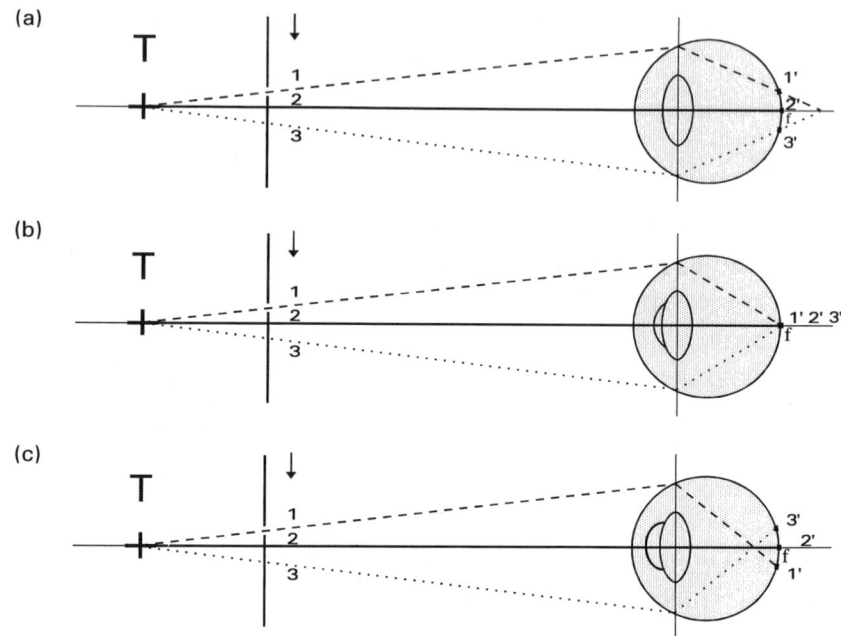

Fig. 6 Retinoscopic principle. Schematic drawing of the side-view of the eye showing the rays of light from the target T through an aperture that is swept through positions 1 (dashed), 2 (solid) and 3 (dotted). **(a)** For under-accommodation, the target images (1′, 2′ and 3′) on the retina move from top to bottom. **(b)** For correct accommodation, the target images on the retina remain stationary. **(c)** For over-accommodation, the target images on the retina move from bottom to top. The phase difference in the signals of photodetectors, located optically conjugate to the retina (not shown), provides a measure of the accommodative response.

Moreover, the electronics can be adopted to provide dynamic accommodative response measurement with a bandwidth of 16 Hz.

Dynamic vergence and saccadic eye movement measurement

Eye movements, or rotations of the eyes, in the horizontal dimension can be measured by illuminating a beam of infrared source that is mounted on spectacles at the limbus of the eye and detecting the amount of reflected light.

The limbus is the boundary between the white sclera and darker iris (as seen through the transparent cornea) portions of the eye. Suppose a detector is placed in front of the lateral limbus (i.e. the boundary between the white sclera and the dark iris) of the right eye. As the eye rotates leftward or rightward, more or less respectively, of the reflected infrared light will be received at the detector, thus providing a quantitative measure of the amount and direction of eye rotation. To reduce low frequency interference, the emitters are pulsed at 1 KHz, and to improve common-mode rejection, two infrared emitter-detector pairs are aimed at the lateral and medial limbus of each eye. The difference between lateral and medial detector signals is then bandpass-filtered at 1 KHz and demodulated to provide a signal proportional to the eye rotation (see Figs. 7A–C). A typical infrared eye movement monitor, such as the Skalar Model 6500, has a linear range of 25 deg, a resolution of 5 min of arc and a bandwidth of 200 Hz. This non-invasive device can be worn comfortably for a relatively long period of time (e.g. 1 to 2 hrs). Since a majority of eye movements occur in the horizontal plane and these movements have a negligible vertical component,[27] one-dimensional horizontal tracking is generally adequate. However, in other applications such as eye fixations on video displays, two-dimensional eye tracking is needed. This can be performed using the Iota Eyetrace Systems (Sweden) eye tracker.

Other eye movement measurement techniques also provide tracking in two dimensions. In the video-based helmet-mounted eye movement monitor system (ISCAN), a miniature infrared camera mounted on the helmet scans the eye. The electronic circuitry tracks the corneal reflection of a distant infrared source and also calculates the center of the pupil. Since the pupil center moves with both eye translation and rotation (whereas the reflection from the spherical corneal surface is influenced by translation but not by rotation), the difference between these two signals provides a measure of eye rotation. The video-based system has a linear range of ±30 deg, resolution of 1 deg and bandwidth of 30 Hz.[106] However, since much higher bandwidths are needed for resolving fine dynamic features of saccade and vergence, it is inadequate for measuring dynamics of eye movements. Instead, the video-based system is more appropriate for assessing two-dimensional fixations. A two-dimensional system that has the needed higher bandwidth uses the scleral search coil technique.[127] In this technique, the subject is enclosed in a large box-shaped frame, which contains circular wound (Helmholtz) coils on either side for the horizontal

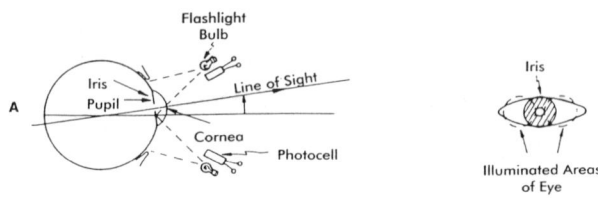

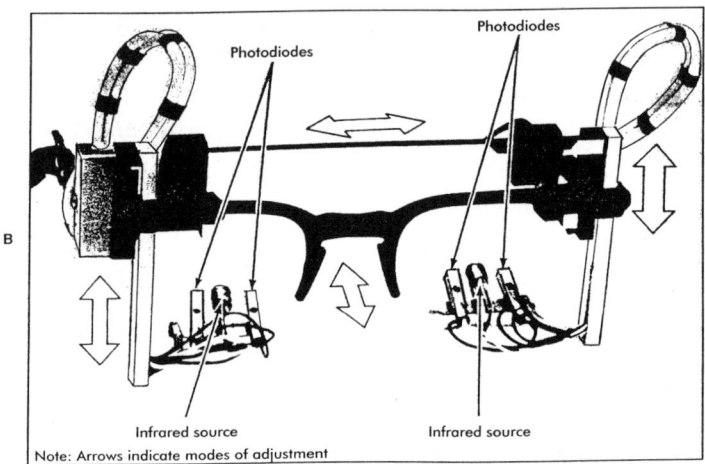

Fig. 7 Infrared limbal eye tracker. **(A)** Light and photocell arrangement for infrared limbal tracking. **(B)** Spectacle-mounted infrared differential reflectivity device (reprinted from Ciuffreda and Tannen,[24] p. 199, with permission).

signal, and on the top and bottom for the vertical signal. The horizontal and vertical coils are driven at 50 and 75 KHz, respectively. The oscillating electromagnetic fields are picked up by a thin circular wire coil embedded in a contact lens worn by the subject. Thus, the eye coil simply acts as a transformer. When the eye coil rotates, there is an induced voltage that is proportional to the sum of the horizontal and vertical magnetic fluxes traversing in the coil. This voltage is amplified and the 50 and 75 KHz components are separated by phase-locked detectors to give the signals for horizontal and vertical components of eye rotation. For rotations of less than 30 deg, the output voltage is linearly related to eye angle to within 5%. The resolution can be as small as 1 sec of arc, and the bandwidth is up to 1 KHz.[24] However, this invasive method requires that the eye be anesthetized and the recording can be performed for only about 20 min. Moreover, the eye coils are delicate and expensive.

For a detailed review of a variety of eye movement measurement devices, see Young and Sheena[171] and Ciuffreda and Tannen.[24]

Static Analysis Techniques

Accommodation System

The accommodation system senses blur of the retinal image and varies the lens power using neurological feedback control to reduce blur to a minimum.[7,158,164] This accommodative ability to focus for objects in depth can be quantified by plotting the accommodative response (AR) as a function of the accommodative stimulus (AS)[110] (see schematic curve in Fig. 8A, and experimental data and simulated curve in Fig. 8B). A simplified descriptive model of the accommodation system, containing various block elements, is shown in Fig. 9A. The difference between the desired optical power (or AS) and the lens response (or AR) provides the blur signal that drives the accommodative controller. The accommodative controller output, in turn, is summed with the tonic input to drive the lens, or plant, to provide the closed-loop AR. The AR is fed back to complete the feedback loop. However, if the feedback loop is disabled, such as under the darkness condition, then the accommodative controller output goes to zero, and only tonic input drives the AR to provide the resting state, or tonic, accommodation.

A more detailed block diagram model of the accommodation system[66] is shown in Fig. 9B. Three elements, the DOF (±DSP), accommodative controller gain (ACG), and tonic accommodation (ABIAS), are of primary importance in determining the characteristics of the static accommodative response. As in other optical systems, the aperture, or pupil, is the primary determinant of the DOF. To model the DOF, an ideal nonlinear static mathematical function, called the deadspace operator, is used. It approximates the threshold of detection of the accommodation system[9] (see Fig. 9B). Thus, if the input to the deadspace operator is greater than the threshold (i.e. +DSP), the output is equal to (input −DSP). On the other hand, if the input is between ±DSP (with threshold, DSP, being a positive number), the output is zero. Further, if the input is less than −threshold (i.e. −DSP), the output is equal to (input +DSP). In other words, the accommodation system needs to sense an error, or blur, to maintain a stable feedback response. Otherwise, without error information, it

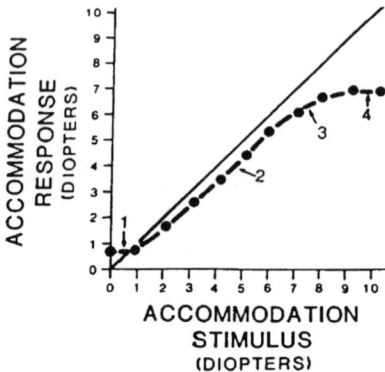

Fig. 8A Schematic static accommodative stimulus-response curve for a typical normal subject. 1 = initial nonlinear portion, 2 = linear region, 3 = transitional soft saturation region, and 4 = hard saturation presbyopic region (reprinted from Hokoda and Ciuffreda,[60] p. 102, with permission).

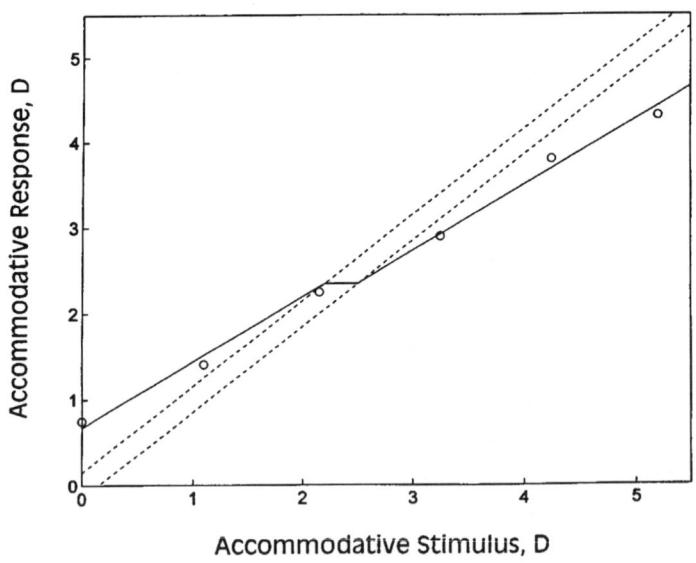

Fig. 8B Typical experimental accommodative stimulus-response data[110] are shown as open circles. The solid curve represents the optimal mean-squared error solution of the accommodation model (reprinted from Hung,[65] p. 335, with permission).

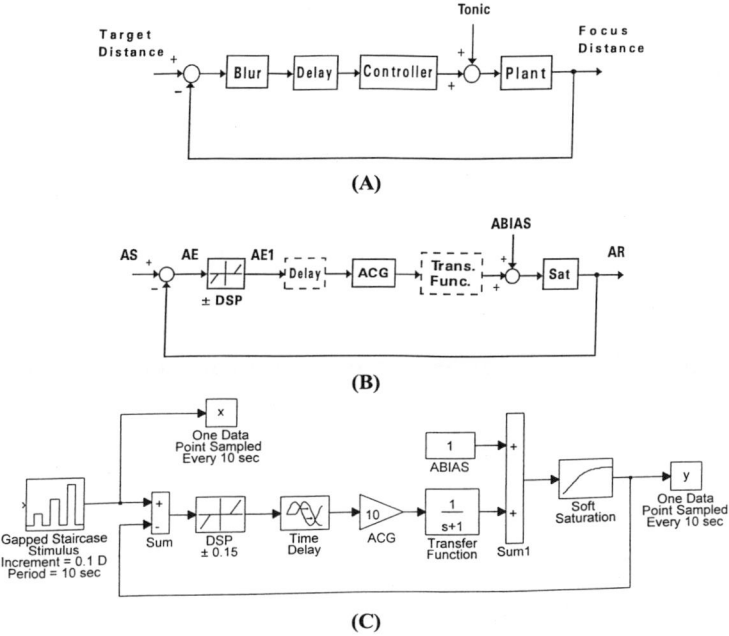

Fig. 9 Accommodation system models. **(A)** Descriptive model. The difference between target distance and focus distance provides the accommodative error, or retinal image blur, that is processed by the accommodative controller following a time delay. The controller output is summed with the tonic signal to drive the accommodative plant or the lens. The feedback loop reduces the blur to a minimum to provide clear focus of the target on the retina. **(B)** Parametric model. The difference between accommodative stimulus and response, AS−AR, provides the accommodative error, AE. The AE is input to the deadspace operator which represents the depth of focus with threshold limits at ±DSP. The output of the deadspace operator, AE1, drives the accommodative controller consisting of gain ACG and a unity-gain dynamic transfer function, following a time delay. The output of the controller is summed with the tonic level, ABIAS, to drive the accommodative plant. The plant is represented by a saturation element, Sat, whose saturation level decreases with age. (The dynamic elements are shown as dashed blocks.) **(C)** MATLAB4.2/SIMULINK1.3 Simulation Model. This is the model used for the simulations. The difference between the gapped-staircase stimulus, with period 10 sec and increment 0.1 D, and the system response is input to the deadspace operator having breakpoints at ±DSP. The deadspace operator output is input to the controller with gain ACG and transfer function $1/(s+1)$ following a time delay. The controller output is summed with the tonic level, ABIAS, to drive the plant, which is represented by a soft saturation element. The steady-state or static levels are obtained by sampling once every 10 sec to give the static stimulus and response functions, x and y, respectively. The static data are plotted for different parameter values (see Figs. 10A–D; reprinted from Hung,[65] p. 336, with permission).

would wander about as if it were open-looped. This would occur if the response is either inside the DOF, where everything appears clear, or when it is well outside of the DOF, where everything appears completely blurred. The accommodation system appears to obtain the necessary error information by oscillating near the edge of the DOF so that, on average, the AR is near the boundary of the DOF. This provides just sufficient blur to maintain stability of the feedback response.

The ACG determines the magnitude of the AR. The higher the gain, the closer is the response to the stimulus (see Eq. 1B). However, this occurs at the expense of increasing dynamic instability. Indeed, due to a latency of about 360 msec, the closed-loop accommodation system would become unstable if the steady-state gain was used in a continuous feedback loop.[72] Thus, Hung and Ciuffreda[72] proposed that the controller consists of an initial fast open-loop component that operates for large errors, followed by a slow closed-loop component that operates for small errors. The overall steady-state gain can be estimated from the slope of the static accommodative stimulus-response curve.[53,110]

ABIAS is a constant accommodative bias level obtained under reduced visual cue conditions, such as in the dark or in an empty field environment.[133,134] The ABIAS value remains about the same for any particular subject under different night and empty field conditions, but varies among different subjects, ranging from about 0.5 to 3.0 D, with a mean of 1.0 D.[79,99,133,134]

The shape of the accommodative stimulus-response curve is influenced by the tonic and DOF components. This is because for smaller stimuli (<1D), ABIAS has a relatively greater influence and drives the response above the stimulus +DSP level (i.e., lead of accommodation). On the other hand for larger stimuli (>1D), ABIAS has a relatively smaller influence so that the response is below the stimulus−DSP level (i.e. lag of accommodation). The overall result is an inflection in the normal accommodative stimulus-response plot (Figs. 8A and B).

The model has been used to quantify AR under a variety of clinical conditions,[17,18,80] and the results have provided important insights into the relationship between the three model parameters as well as the development of ocular anomalies such as nystagmus, amblyopia, strabismus and myopia. Recently, a systematic analysis of the effect of parameter variation on the accommodative stimulus-response function was performed. The details of the model development are described elsewhere[66] and is summarized below.

The AR operates on either side of the deadspace region about the 1:1 line (a term used in the clinic to indicate a theoretical line that provides a one-to-one relationship between stimulus and response). The equation for the lower deadspace line (see Fig. 8B, lower line) is given by

$$AR_{ld} = AS - DSP. \qquad (2)$$

The equation for the upper deadspace line is given by

$$AR_{ud} = AS + DSP. \qquad (3)$$

It can be shown that for larger AS values, AR is below AR_{ld}, but for smaller AS values, AR is above AR_{ud}. The crossover occurs at

$$AR = ABIAS \qquad (4)$$

It can be shown that for operation on the low side of the deadspace operator, where the response "lags" the stimulus, the AR is given by

$$AR = \frac{ACG}{1+ACG} \bullet AS + \left[-\frac{ACG}{1+ACG} \bullet DSP + \frac{1}{1+ACG} \bullet ABIAS \right]. \qquad (5)$$

The model shown in Fig. 9C was simulated using MATLAB4.2/Simulink1.3. A novel simulation technique was used to provide a relatively simple means to obtain the steady-state characteristics of the nonlinear static model of the accommodation system. The stimulus consisted of a gapped staircase pattern having a period of 10 sec, so that the zero-value duration as well as the step duration were both equal to 5 sec. This step duration was long enough for the transients to dissipate and reach a static steady state value. This was verified in control tests by comparing responses at 5 sec and at 100 sec, which showed no difference in their final values. The increment for each subsequent step was set at 0.1 D. Thus, for a simulation duration of 500 msec at a period of 10 sec, there were 50 input and 50 output static points, and a total static stimulus range of 5 D.

Four parameters were varied: the deadspace range, ±DSP; the accommodative controller gain, ACG; the tonic level, ABIAS; and the simulated age, which was represented by the plant saturation level. A soft saturation element was used to simulate plant saturation. In the model simulation of AR as a function of age, the ages 20, 30, 40 and 50 corresponded to plant saturation levels of 10, 5.9, 3.0 and 1.1 D, respectively.[33,52,59]

In addition to the above, the model parameters were fitted to the data (open circles) of Fig. 8B using an optimization procedure[66] (see solid curve, Fig. 8B). First, equations were written for the mean-square error between the model and the data in the form

$$f(k) \sum_{k=1}^{6} [M(k) - D(k)]^2 \qquad (6)$$

where $M(k)$ and $D(k)$ were the model equations and experimental data, respectively, for the $k=6$ data points. The "minimax" command in MATLAB4.2/SIMULINK1.3, which minimized the maximum error of $f(k)$ in Eq. 6, was used under the constraints $0.15 \leq DSP \leq 0.30$, $0 \leq ACG \leq 10$ and $0 \leq ABIAS \leq 4$, with the initial conditions DSP=0.20, ACG=5 and ABIAS=2.

The model sensitivity-analysis results showed that increasing the deadspace operator breakpoints, ±DSP, increases the separation between the solution lines (Fig. 10A). Within the deadspace region, the response (i.e. the horizontal line segment) is equal to ABIAS. For stimuli greater than ABIAS+DSP, the AR is below the lower solution line. This response line begins with a value of ABIAS at the stimulus level equal to ABIAS+DSP, and has a slope equal to ACG/(1+ACG). However, for stimuli less than ABIAS−DSP, the AR is above the upper solution line, and begins on the upper solution line with a value of ABIAS at the stimulus level equal to ABIAS+DSP, and has a slope equal to ACG/(1+ACG). Figure 10A shows that as the DOF increases, the model AR curve shifts away from the 1:1 line, while maintaining the same slope. Such a shift in the curve has been observed experimentally in subjects with peripheral optical and/or retinal deficits, and in late-onset myopia (i.e. development of myopia, or nearsightedness, after age 15 years). For example, Ong et al.[118] found that increased DOF was the primary contributor to the anomalous accommodative behavior in individuals with congenital nystagmus (a regular, repetitive, and usually rapid involuntary movement or rotation of the eye that is either oscillatory or has slow and fast phases in alternate directions), with this perhaps being related to abnormal fixational eye movements and eccentric fixation. Also, Jiang[86] found that late-onset myopes exhibited a relatively large shift in their experimental accommodative stimulus-response curves. This may be due to long-term exposure to changes in nearwork-induced retinal defocus that led to myopia development (see Refractive Error Development Model section), with the subsequent continued exposure resulting in degraded AR.

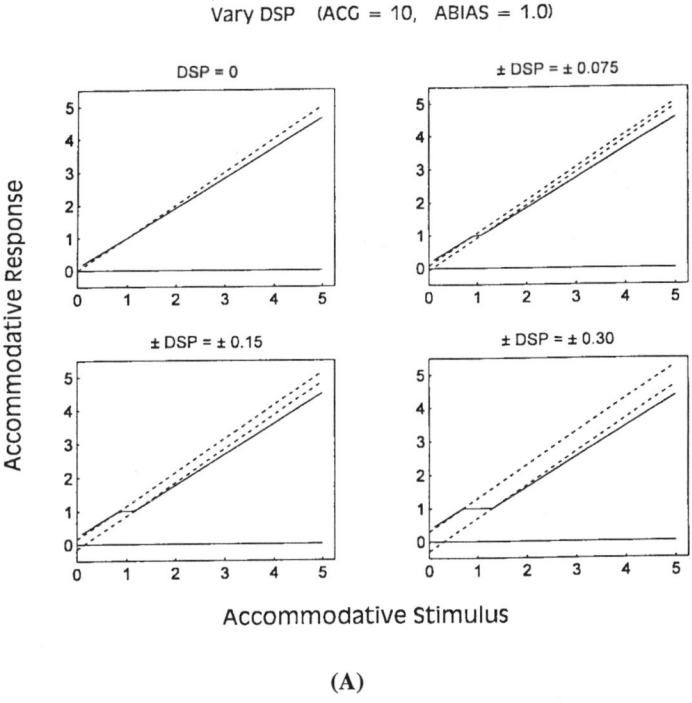

(A)

Fig. 10 Sensitivity of the model accommodative stimulus-response relationship (see Fig. 9C) to parameter variations in **(A)** DSP, **(B)** ABIAS, **(C)** ACG, and **(D)** Age (nominal values: DSP = ± 0.15 D, ABIAS = 1.0 D, ACG = 10, Age ≤ 30 yrs). For all four sub-plots, the dashed lines represent the limits of the deadspace (±DSP) and the solid lines represent the simulation results. The stimulus-response curve to the right of the deadspace lines begins at the response level equal to ABIAS. The slope is equal to ACG/(1 + ACG). Similarly, the curve to the left of the deadspace lines begins at a response level equal to ABIAS and has the same slope (reprinted from Hung,[65] p. 338, with permission).

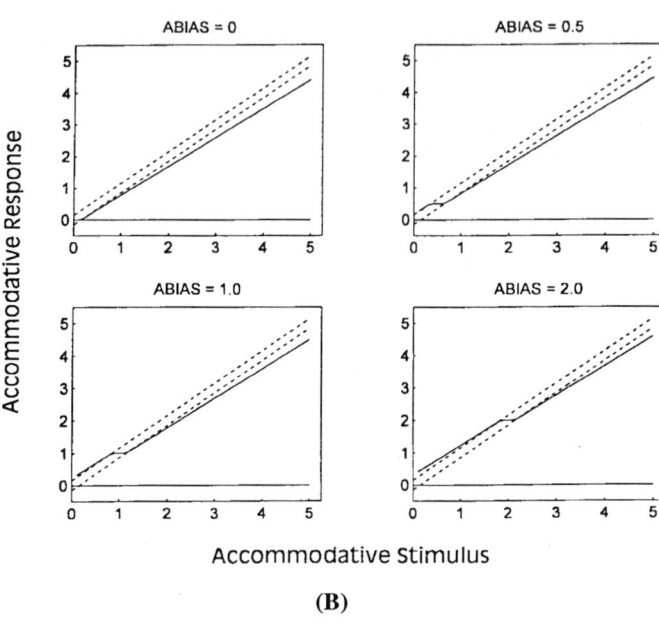

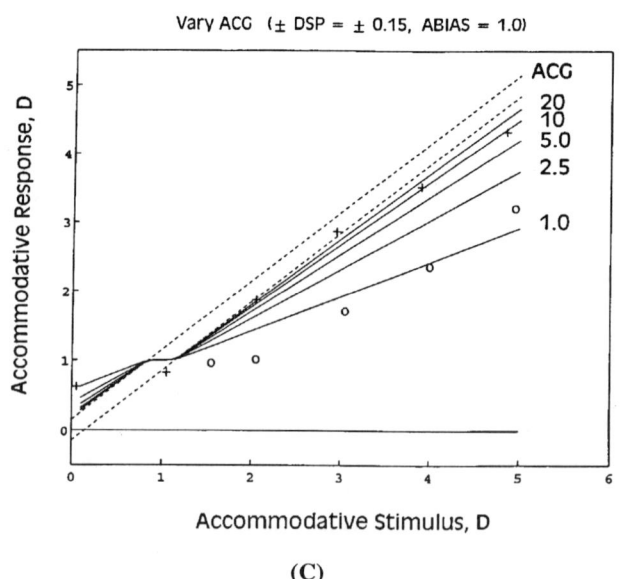

Fig. 10 (Continued)

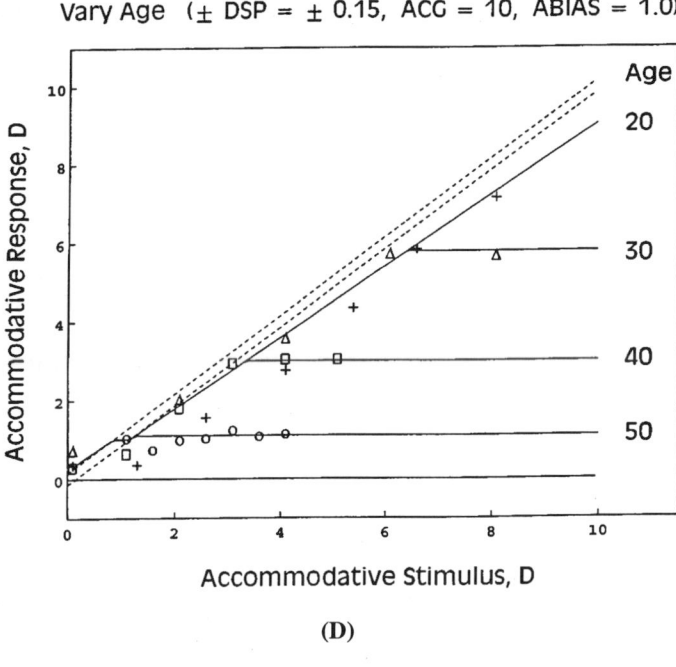

Fig. 10 (Continued)

Increasing the tonic level, ABIAS, shifts the level within the deadspace but maintains the same relative positions of the AR lines outside the deadspace region (Fig. 10B). The shifts in ABIAS level and the associated changes in the stimulus-response curve has been observed experimentally.[17] Under darkness or open-loop conditions, the controller output is zero, and only ABIAS drives the AR. This can be seen in the model in Fig. 9B. However, although this tonic value is relatively large under darkness or open-loop conditions, it has only a relatively small influence on the AR under normal closed-loop conditions.[77,133,134] The inflection of the accommodative stimulus-response curve at ABIAS provides a relatively simple means to obtain an estimate of the ABIAS level from the closed-loop accommodative stimulus-response data (see Fig. 10B). However, such a transition may not be easily observed experimentally. When this is the case, an optimal simulation such as that illustrated in Fig. 8B (solid curve), and discussed below, can be used to determine the level where the inflection occurs, and thus provide an estimate of the ABIAS level. In the experimental AS/R curve shown in Fig. 8B, the inflection occurs at 2.5 D. Therefore, for this subject, the ABIAS level would be equal to 2.5 D. Although the ABIAS level was not measured in the study[110] used for Fig. 8B, evaluation of previous data[77] where both the AR and ABIAS levels were measured, confirms this relationship. The intersection of the linear regression line and the 1:1 line is called the crossover point. It had been used by some researchers[12] to estimate the ABIAS level. However, the AR at the crossover point is typically about 1.5 D, which is above the 1 D ABIAS value obtained experimentally.[133,134] Figure 8B illustrates how such an overestimation could have occurred (see Hung[66]). Essentially, if the nonlinear transition or inflection region is not appropriately taken into account, a linear regression line drawn through all the data points would intersect the 1:1 line above the transition region, or ABIAS level. The only case when the crossover point is at the ABIAS level is if the DOF or ±DSP is zero, so that the dashed curves in Fig. 8B collapse into the 1:1 line. However, since the DOF is generally nonzero under normal viewing conditions, the crossover point using a regression line through all the data points would over-estimate the ABIAS value.

Increasing the ACG increases the slope of the AR, with the simulated slope equal to ACG/(1+ACG). Note that the inflection of the curve occurs at an AR level equal to ABIAS for all the different ACG curves. Data for normal and amblyopic eyes[17,21] are also shown in Fig. 10C. Clinically, amblyopia is said to exist if vision is 20/30 or worse. The normal ACG ($\approx$ 10) is maintained by central neural controller signals[77] (Fig. 10C). However, defects in the afferent (e.g. decreased contrast sensitivity[20]) and/or efferent (e.g. decreased neural command signal) pathways would result in a reduction in the controller gain, and in turn the slope of the accommodative stimulus-response curve. Thus, the reduction in slope, and therefore ACG, seen in the amblyopic eye (Fig. 10C) indicates a central, rather than peripheral neural deficit.[21] Indeed, the value of ACG has been found to be a useful indicator of amblyopic deficit.[17,21] Moreover, some experimental data show both a shift away from the 1:1 line and a reduction in slope (not shown, see Ciuffreda et al.[17]). This indicates that in these subjects, where DSP as well as ACG are affected, the deficit could be attributed to both peripheral and central factors.

With increasing age, the saturation level decreases, while maintaining the same normal AR for values below the saturation level (Fig. 10D).

Representative experimental data for the different age groups[22,77,108,128] are also shown in Fig. 10D. The model accurately simulated the experimental AR as a function of age (Fig. 10D). The data show a decrease in saturation level as age is increased. However, the slope of the linear portion remains the same. Two prominent theories of presbyopia, or the decline in accommodative amplitude with age, have been proposed previously to account for the experimental data. In the Hess-Gullstrand theory, the lens and capsule harden with age, thus reducing the maximum level of accommodation.[57,159] Such a saturation effect would be reflected in a shape of the stimulus-response curve that looks like those of Fig. 10D. Thus, this theory is compatible with experimental results.[108,124] On the other hand, in the Duane-Fincham theory, the ciliary muscle weakens with age.[32,159] The shape of the stimulus-response curve would look like that of Fig. 10C for changes in ACG. The theory is based on studies in which the administration of dilute atropine solution, which relaxes the ciliary muscle, results in a decline in gain of the accommodative stimulus-response function. However, experimental evidence indicates that, in the absence of drugs, the contractile capability of the ciliary muscle remains unaffected with increasing age.[41,141] Thus, the Hess-Gullstrand theory provides the more reasonable explanation for the decline in accommodative amplitude with age. This is represented in the model by a decrease in the saturation level of the soft saturation element of the plant.

While some of the oculomotor parameters remain unchanged, other oculomotor parameters have been found to vary with age.[23,108] The objective DOF (i.e. objective instrument measurement of overall deadspace range) was found to remain constant with age. This indicates that the fundamental mechanism of threshold blur detection and the associated neurologically-based accommodative reflex remain unchanged with age. On the other hand, the subjective DOF (i.e. the subjective assessment of overall deadspace range) increases at a rate of 0.027 D/yr. This can be attributed to an increased tolerance of the error in defocus blur as one ages, despite the ability to respond to these threshold stimuli. The ACG remains unchanged with age. This indicates the accommodative neural controller signal remains normal and constant with age. ABIAS decreases with age at a rate of -0.04 D/yr. This is not attributed to any changes in the neural controller signals that had maintained the tonic level at a younger age. Rather, it is due to the progressive loss in motor responsivity as a result of age-related decrease in the saturation level of the crystalline lens and related peripheral structures (i.e. the plant) that participate in the accommodative process. These findings, therefore, provide valuable information regarding model parameter changes with age, and can serve as a quantitative basis for clinical assessment of AR as a function of age.

Generally, a single parameter had been previously used to characterize the entire accommodative stimulus-response function. A number of investigators used the slope of the stimulus-response curve.[87,104] However, because the nonlinear deadspace operator with boundaries given by AR = AS ± DSP (see dashed lines in Fig. 8B), divides the response into three regions, a single overall slope measure covering all three regions would not accurately reflect the characteristics of the response function. As shown above in the simulation results, the slope in the region at the inflection region is equal to zero. Thus, the single overall calculated slope would be somewhat lower than the actual slope in the linear region. Chauhan and Charman[13] introduced the concept of a single index for the stimulus-response function based on the area between the best-fit curve and the 1:1 line over the stimulus interval between 1.5 and 4.5 D. The accommodative error index is obtained by dividing the area by the stimulus interval. Thus, it is essentially the mean-squared error between the response curve and the unity ratio line. One problem with this measure is that it does not take into account the DOF, whose lower boundary delimits the largest AR possible under the physical optical constraints. Indeed, under monocular viewing conditions and in the absence of crosslink influence from vergence,[77] the AR has been shown to operate up to the boundary of the DOF.[81] Hence the calculated error (between the 1:1 line and the response) would be larger than the actual error (between the lower DOF boundary line and the response).[70] In contrast to these single parameter approaches, the three-parameter model presented here provides a consistent and accurate quantification of the nonlinear characteristics of the accommodation system. In the absence of physiological measurements of the DOF (± DSP) and ABIAS, these parameters can be estimated. A reasonable estimate for DSP is 0.15 D.[77] ABIAS can be estimated from the accommodative level at the inflection of the optimal simulation curve or from the darkness experiment (see Fig. 8B). The third parameter, ACG, can be calculated from the appropriate slope of the stimulus-response function to the right of the crossover point. Alternatively, ACG can be obtained from Eq. 5 to give ACG = (AR − ABIAS)/(AE − DSP),[77] and data values can be substituted into the equation to obtain an average estimate of ACG.

The optimization procedure presented here (see Eq. 6 and the constraint equations) can be applied relatively easily using the MATLAB4.2/SIMULINK1.3 Optimization Tool. Thus, the accommodative stimulus-response data over the entire stimulus range can be directly input into the computer to provide optimal calculated values of the model parameters. Because of widespread availability of MATLAB software, this procedure can serve as a standard for obtaining the model parameter values of the accommodation system.

Vergence System

In contrast to the accommodation system, which exhibits relatively large errors typically of about 0.3 to 0.5 D,[116] the vergence system is relatively accurate, where the error is typically less than 5 min of arc. Because the static vergence response is usually very close to the stimulus, it is usually not directly plotted against the stimulus. Instead, the difference between the vergence stimulus (VS) and response (VR), called vergence error (VE) or fixation disparity (FD), is usually plotted as a function of the VS. The effect of VS on FD provides a measure of the gain of the vergence system[77] (see Eq. 1B), with a smaller FD corresponding to higher gain.

Also, the vergence system static response is generally not studied in isolation. This is because there is an important interactive link between accommodation and vergence. The accommodative convergence crosslink provides a quantitative measure that is useful for the assessment of the "strength" of the drive from accommodation to vergence, which in turn provides a clinical indicator of the integrity of the patient's oculomotor balance.

Linear Analysis of Relationship Between AC and ACG

The coordination of binocular visual fixation in depth, along with accommodation in each of the two eyes, are two of the most important functions of the eye movement system. It provides clear focus and single binocular vision to assess accurately targets in depth, which has important early phylogenetic implications on survival. As discussed earlier, accommodation is driven by blur of the target image on the retina. On the other hand, vergence or rotation of the two eyes in opposite directions, is driven by the disparity of the images on the retinas.[39,45,167] There is an inherent linkage between accommodation and vergence,[101,110] and the two systems form an interactive dual-feedback system.[77,78] For example, when one eye is blocked to remove disparity cues, changing focus of the viewing eye results in a rotation of the eye under cover that is in direct proportion to the change in focus. This proportionality factor is called the accommodative convergence to accommodation (AC/A) ratio.[1,19] Conversely, when the two eyes view a target through pinhole apertures to remove blur cues, convergence results in a change in focus that is in direct proportion to the amount of convergence. This proportionality factor is called the convergence accommodation to convergence (CA/C) ratio. This tight linkage between these two sub-systems provides a coordinated oculomotor response to target changes in depth. On the other hand, disturbance in this linkage may result in oculomotor disorders such as strabismus (or crossed-eye) and convergence insufficiency. In the clinic, the AC/A ratio is measured to provide an assessment of the linkage between accommodation and vergence.[5]

The strength of the accommodative convergence drive can be measured in two ways. When the vergence output is measured relative to the accommodative *response*, it is called the accommodative convergence to accommodative response ($AC/A_{response}$) ratio. On the other hand, when measured relative to the accommodative *stimulus*, it is called the accommodative convergence to accommodative stimulus ($AC/A_{stimulus}$) ratio. Alpern[1] noted from experimental results that the ratio of these terms, $R = (AC/A_{response})/(AC/A_{stimulus})$, is about 1.08 in visually normal individuals. The reason for deriving R is that $AC/A_{response}$ is more difficult to measure than $AC/A_{stimulus}$. Thus, usually in the clinic, the $AC/A_{stimulus}$ ratio is measured, and then the $AC/A_{response}$ ratio can be obtained by multiplying by 1.08. However, this is not done for patients with abnormalities, such as in divergence excess, where the ratio can be as high as 1.30.

To determine the rationale for the value of R being equal to 1.08, a quantitative analysis was performed on the accommodation and vergence model with vergence open-loop (Fig. 11A). If the AR lags the stimulus, as is usually the case when vergence is open-looped, the deadspace operator can be replaced by a bias term (Fig. 11B). This results in a linear model which can be analyzed systematically. Consider Eq. 5, if two measurements are made for near and far accommodative stimuli, respectively, we obtain

$$AR_{near} = \frac{ACG}{1+ACG} \cdot AS_{near}$$
$$+ \left[-\frac{ACG}{1+ACG} \cdot DSP + \frac{1}{1+ACG} \cdot ABIAS \right] \quad (7)$$

$$AR_{far} = \frac{ACG}{1+ACG} \cdot AS_{far}$$
$$+ \left[-\frac{ACG}{1+ACG} \cdot DSP + \frac{1}{1+ACG} \cdot ABIAS \right]. \quad (8)$$

If the corresponding open-loop vergence responses are VR_{near} and VR_{far}, respectively, then by combining with Eqs. 7 and 8, we obtain the response AC/A:

$$AC/A_{response} = \frac{VR_{near} - VR_{far}}{AR_{near} - AR_{far}} = \frac{VR_{near} - VR_{far}}{\frac{ACG}{1+ACG} \cdot (AR_{near} - AR_{far})}. \quad (9)$$

On the other hand, the stimulus AC/A is given by

$$AC/A_{stimulus} = \frac{VR_{near} - VR_{far}}{AR_{near} - AR_{far}}. \quad (10)$$

Now dividing Eq. 9 by Eq. 10 gives the ratio of response to stimulus

$$R = \frac{AC/A_{response}}{AC/A_{stimulus}} = \frac{1+ACG}{ACG}. \quad (11)$$

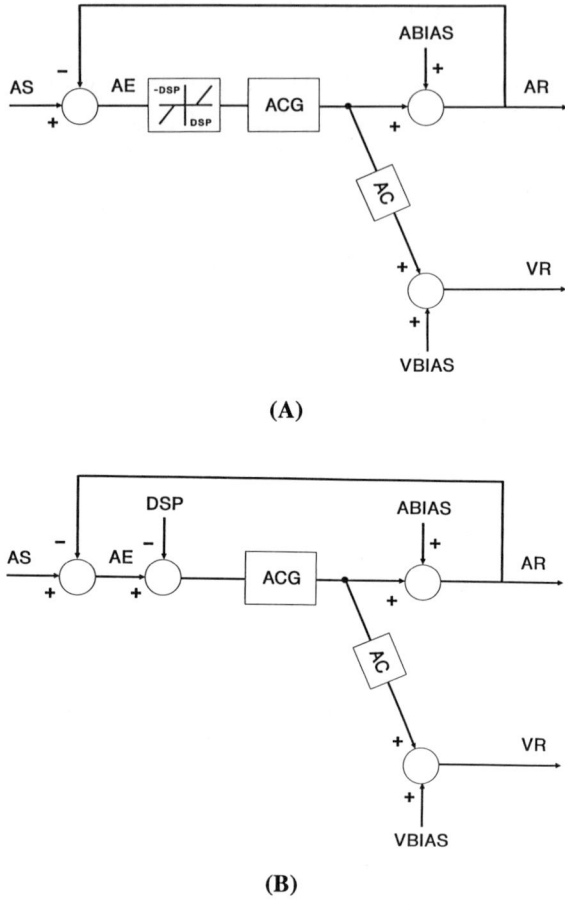

Fig. 11 (A) Linear static model of interactive dual-feedback accommodation and vergence system, with vergence system open-looped. Accommodative error (AE), or blur, is the difference between accommodative stimulus (AS) and accommodative response (AR). It is input to a deadspace operator, ±DSP, representing the depth of field. The output of the deadspace operator is multiplied with the accommodative controller gain, ACG, to give the accommodative controller output. The accommodative controller output is summed with the accommodative bias (ABIAS), or tonic accommodation, to give the accommodative response, AR. It is also cross-linked to the vergence subsystem via gain AC, which is then summed with the vergence bias (VBIAS), or tonic vergence, to give the accommodative convergence response, VR. (B) If the accomodative response lags the accommodative stimulus, as is usually the case when vergence is open-looped, then it can be shown that the deadspace operator can be replaced by a bias term. The other parts of the model remain the same as in (A) (reprinted from Hung,[68] p. 107, with permission).

This is a surprisingly simple and interesting result. For example, using a typical subject ACG value[75] of 10 gives an R = 1.10. This is close to the 1.08 found experimentally. Moreover, the ratio R is dependent on the ACG but not on the crosslink gain AC. Thus, this quantitative analysis was able to provide insight into the relationship between the model parameters and the response and stimulus AC/A ratios.

Nonlinear Analysis of AC/A Using the Phoria and Fixation Disparity Methods

In the 1960s, well before the development of the quantitative interactive dual-feedback model of accommodation and vergence,[77] Ogle[114] attempted to exploit the close interaction between accommodation and vergence by proposing a measure of oculomotor linkage that is obtained under the more natural binocular viewing condition. Instead of the conventional AC/A ratio measured under monocular viewing, called the phoria method, they proposed that the AC/A ratio be obtained under fused binocular viewing, called the FD method. For a particular viewing distance, the FD was measured for different prism-induced VS, while the AS was held constant. Then, the FD was measured for different lens-induced AS while the VS was held constant. The viewing distance was typically fixed at 40 cm for both lens and prism viewing. First, the AS was held constant at 2.5 D, and the prism was varied over a range of base-out (BO) and base-in (BI) levels. For each prism stimulus level, the FD was measured subjectively using a device such as the Sheedy disparometer.[151] During each experimental trial, the subject viewed the disparometer through cross-polarized spectacles so that the right eye saw only the upper, while the left eye saw only the lower vertical line segment. The horizontal separation between the two vertical line segments was varied until the subject determined that they were perceptually aligned. The physical separation of the lines provided the measurement of FD. Then second, the VS was held constant at 2.5 MA, and the lens was varied over a range of dioptric levels. For each lens stimulus level, the FD was similarly measured. Finally, the results were plotted as FD versus prism and FD versus lens curves. For a particular FD, the prism value was matched with the lens value. This was done over a range of FD values, and the resulting data were plotted as prism versus lens values. Ogle et al.[114] proposed that the slope of this prism versus lens plot should be equal to the AC/A ratio obtained under the monocular viewing condition (Figs. 12A–C). However, a comparison of experimental AC/A ratios for 13 subjects[114] showed a difference between the values for FD and phoria methods ranging from −63% to 156%. They

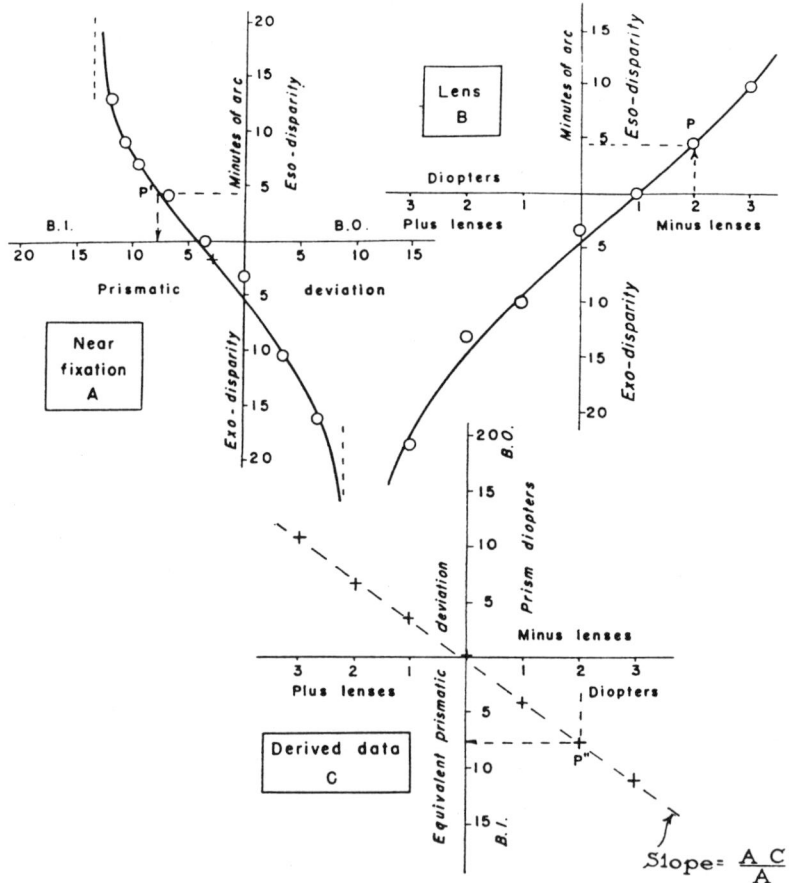

Fig. 12 Graphical representation of the fixation disparity as influenced by **(A)** prisms and **(B)** lenses. Graph C represents the derived data relating the prism power and the lens power that caused the same fixation disparity, where the slope was expected to be equal to the AC/A ratio obtained monocularly. This did not turn out to be the case[65,114] (adapted from Ogle et al.,[114] p. 68, with permission).

suggested that part of the variation may be due to measurement error. But this could only account for about 8% of the difference.[114] Such a discrepancy remained an unresolved puzzle to optometrists and vision scientists for over three decades until the recent comprehensive quantitative analysis of the dual-interactive model.[64,65]

It turned out that although Ogle *et al.*[114] did not use control system analysis, their idea based on matching the AC/A ratios under open- and closed-loop conditions, was insightful. Indeed, if the interactive accommodation and vergence systems were linear (i.e. without the nonlinear deadspace operators DOF and PFA), their method would have yielded accurate matches between the two measures. However, as a comprehensive quantitative analysis of the nonlinear model showed (see below), the deadspace operators produced multiple solutions under the FD method.[65] Since the prism data were obtained on different solution lines for different lens values, the calculated AC/A ratios would be different from that obtained using the phoria method. Thus, as detailed below, the static nonlinear model could account for the discrepancies found between the AC/A ratios using the two methods.

The static nonlinear model was based on a previously developed static model of the accommodation and vergence system.[77] The DOF and PFA, representing the accommodative and vergence deadspace elements, were fixed at ± 0.15 D and ± 6 min of arc respectively, to simulate the normal fixation conditions.[64] Since each of these deadspace operators had two solutions, there were a total of four basic solution equations relating FD to the AS and VS. However, since there were two conditions, prism viewing and lens viewing, this resulted in eight solution equations. In addition, since FD was matched between the prism and lens viewing conditions, the difference between the two sets of solution equations resulted in 16 different solutions. Experimentally derived parameter values for two of the subjects in Table 2 were applied to the model equations.

Derivation of model equations

Open-loop vergence

This corresponds to the experimental condition of monocular viewing in which disparity feedback in the vergence system is rendered ineffective. Consider

Table 2 Control and Adaptation Model Parameters for Four Subjects.
(Adapted from Hung and Semmlow,[77] p. 446, and Hung,[68] p. 322, with permission.)

Parameter	Values for Subjects			
	JS	GH	FR	JM
ABIAS (D)	0.42	1.70	0.75	2.20
VBIAS (MA)	−0.62	1.00	0.19	0.93
ACG	8.53	2.60	10.7	2.32
VCG	293.0	141.0	126.0	166.0
AC (D/MA)	1.15	1.76	0.49	0.63
CA (MA/D)	0.66	0.50	0.85	0.47
K_A	4	10	12	10
T_{A1} (s)	25	25	25	25
T_{A2} (s)	4	4	4	4
m_A	3	0.5	0.5	0.5
K_V	9	6	6	5
T_{V1} (s)	50	50	50	50
T_{V2} (s)	8	8	8	8
m_V	0.5	2	2	2

the model shown in Fig. 13. If the vergence system is open-loop, the output of the CA cross-link component would be zero. It can be shown that under this condition the VR is given by:[65]

$$VR = \frac{ACG}{1+ACG} \cdot AC \cdot AS$$
$$+ \left[VBIAS - \frac{ACG}{1+ACG} \cdot (ABIAS + AD) \cdot AC \right]. \quad (12)$$

Let VR_1 and VR_2 be the accommodative vergence responses to accommodative stimuli AS_1 and AS_2, respectively. Substituting each of the two conditions into Eq. 12 and subtracting, we find that the terms within the square brackets cancel. Rearranging the equation, we obtain the stimulus AC/A ratio or

$$AC/A_{stimulus} = \frac{VR_2 - VR_1}{AS_2 - AS_1} = \frac{ACG}{1+ACG} \bullet AC. \quad (13)$$

It can be shown[78] that in the linear AS/R range, $AC/A_{response} = AC$. Thus, this nonlinear model solution under the vergence open-loop condition is the same as that for the linear model[65] (also see Eq. 11). The equality between the linear and nonlinear models is expected since with vergence open-loop, the accommodation system essentially operates on only one side of the deadspace element, and thus can be represented by a simple bias term. The deadspace limit value, AD, is embedded in the bracketed term in Eq. 12. Taking the difference between two readings eliminates the bracketed term in Eq. 12. Thus, the resulting equation for AC/A (Eq. 13) is the same as that for the linear model.[65]

Closed-loop vergence

This corresponds to the experimental condition of binocular viewing of a target at a fixed distance while either the prism or the lens is varied. Thus, in the

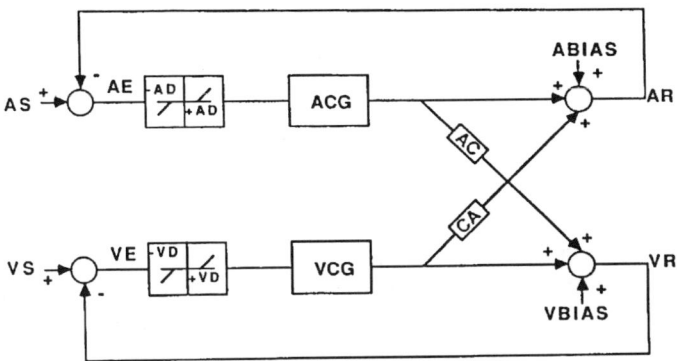

Fig. 13 Nonlinear static model of interactive dual-feedback accommodation and vergence system containing the deadspace operators depth of field (DOF) and Panum's fusional area (PFA). The deadspace between + and − AD simulates the DOF (AD = 0.15 D). The output of the deadspace operator, AE ± AD, is multiplied with accommodative controller gain, ACG, to give the accomodative controller output. The deadspace between + and − VD simulates PFA (VD = 6 min of arc). The output of the deadpsce operator, VE ± VD, is multiplied with vergence controller gain, VCG, to give the vergence controller output (reprinted from Hung,[65] p. 308, with permission).

model, the accommodation and vergence sub-systems combine to form an interactive dual-feedback control system.[77] Each sub-system contains a deadspace operator: DOF for accommodation and PFA for vergence. Stability of static closed-loop accommodation and vergence responses depends on nonzero signals from the outputs of the respective deadspace operators.[64-70] The output can be on either side of the deadspace operator depending on the relative stimulus values to accommodation and vergence.[64] Since each deadspace operator can have two stable solutions, a general static model would contain four combinations of deadspace operator outputs (see Table 3). The model can be analyzed by writing general equations that apply to all the different combinations. It can be shown that the vergence error is given by:[65]

$$VE = \frac{[(1+ACG) \cdot (VS \mp VD \cdot VCG \mp AD \cdot ACG \cdot AC - VBIAS) + (-AS \pm AD \cdot ACG \pm VD \cdot VCG \cdot CA + ABIAS) \cdot ACG \cdot AC]}{(1+ACG) \cdot (1+VCG) - ACG \cdot VCG \cdot AC \cdot CA}. \quad (14)$$

Consider the two stimulus conditions used by Ogle *et al.*[114] In the first condition, the target is at a constant distance, and the prism power is varied. This is represented by

$$AS = AS_c, \text{ and } VS = VS_c + P \quad (15)$$

where AS_c and VS_c are the accommodative and vergence stimuli, respectively, at the fixation distance. The prism power, P, is positive for base-out and negative for base-in prisms.

In the second condition, the target is at a constant distance, and the lens power is varied. This is represented by

$$AS = AS_c - L \text{ and } VS = VS_c \quad (16)$$

where L is the lens power, with minus lens power providing positive stimulus to accommodation.

Substituting the stimulus definitions of Eq. 15 into Eq. 14 gives the FD equation for the *prism viewing condition*, VE_P, which has the same form as Eq. 14. There are four solutions corresponding to the four conditions of deadspace operator output (see Table 3). These are designated as $VE1_P$,

Table 3 Four Combinations of Outputs of Deadspace Operators (AD > 0 and VD > 0).
(As illustrated in Fig. 13; adapted from Hung,[65] p. 308, with permission.)

Combination	Linearized Deadspace Equation		Condition
	AE ± AD	VE ± VD	
(1)	+	+	AE<–AD, VE<–VD
(2)	–	+	AE >AD, VE<–VD
(3)	+	–	AE<–AD, VE>VD
(4)	–	–	AE>AD, VE>VD

$VE2_P$, $VE3_P$ and $VE4_P$. Hence for example, $VE1_P$ is the FD equation for prism viewing under condition 1 in Table 3.

Similarly, substituting Eq. 16 into Eq. 14 gives for the *lens viewing condition*, VE_L, which has the same form as Eq. 14. The four solutions are designated as $VE1_L$, $VE2_L$, $VE3_L$ and $VE4_L$. Hence for example, $VE1_L$ is the FD equation for lens viewing under condition 1 in Table 3.

For the FD technique, the vergence error under the prism and lens viewing conditions are set to equal values. To do this, we first form the difference between VE_P and VE_L, and then *set the difference to zero*. However, there are 16 combinations of differences corresponding to the four prism and lens viewing FD equations. Therefore, we have

$$VE_P - VE_L = (1 + ACG) \cdot P - ACG \cdot AC \cdot L - \begin{bmatrix} \text{nonlinear term,} \\ \text{16 combinations} \end{bmatrix} = 0. \quad (17)$$

Dividing by $1 + ACG$, we obtain

$$P = \frac{ACG \cdot AC}{1 + ACG} \cdot L + \frac{\begin{bmatrix} \text{nonlinear term,} \\ \text{16 combinations} \end{bmatrix}}{1 + ACG} \quad (18)$$

or

$$P = \frac{ACG \cdot AC}{1 + ACG} \cdot L + B_{\text{nonlinear}} \quad (19)$$

46 Oculomotor Control Models

The 16 combinations of the intercept of Eq. 19, $B_{nonlinear}$, can be found in Hung.[65] As in the linear model,[65] we take the difference between two readings and obtain

$$(P_2 - P_1) = \frac{ACG \cdot AC}{1 + ACG} \cdot (L_2 - L_1) + (B_{nonlinear_2} - B_{nonlinear_1}).$$

(20)

Thus, the AC/A ratio for the nonlinear model is given by

$$AC/A_{nonlinear} = \frac{P_2 - P_1}{L_2 - L_1} = \frac{ACG}{1 + ACG} \cdot AC + \frac{(B_{nonlinear_2} - B_{nonlinear_1})}{L_2 - L_1}.$$

(21)

Note that the first term on the right-hand-side is the same as the equations for the linear model[65] and the nonlinear model with vergence open-loop (Eq. 13), but the multiple solutions in the second nonlinear term results in variations in the calculation of the AC/A ratio.

Model simulations

Parameters calculated from experimental measurements for subject GH[77] were input into equations for VE_P and VE_L (see Eq. 14).[65] The results are shown in Fig. 14 for the four conditions of Table 3, where the plot is in the same form as that used by Ogle et al.[114]

The parameters for subject GH were then input to the intercept term, $B_{nonlinear}$ of Eq. 19, for the 16 combinations of the difference $VE_P - VE_L$. The results are shown in Fig. 15. The line through the origin corresponds to: $B_{nonlinear} = 0$ under the vergence closed-loop condition of the nonlinear model; the vergence open-loop solution of the nonlinear model (Eq. 13); as well as the vergence open- and closed-loop solutions of the linear model.[65] The other lines with nonzero intercepts, $B_{nonlinear}$, represent other solutions of the nonlinear model. Note that there are fewer than 16 distinct lines in the graph because of the non-uniqueness of some of the solutions. It could be seen that some of the solution lines were relatively widely separated on the prism versus lens plot. This could explain how the responses could fall on different solution lines, giving calculated slopes under the FD method that were very

Static Analysis Techniques 47

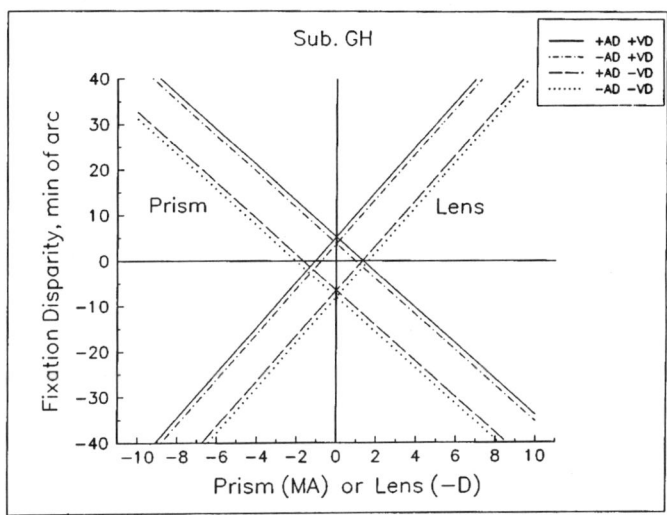

Fig. 14 Model simulation results[65] obtained by inputing parameter values for subject GH into equations for VE_P and VE_L (see Eq. 14). The form of the plot of fixation disparity versus prism or lens is the same as that used by Ogle *et al*.[114] Note that minus lens (towards the right on the abscissa) represents positive stimulus to accommodation. The four solutions for prism viewing (and the four solutions for lens viewing) correspond to the four conditions of Table 3 (reprinted form Hung,[65] p. 308, with permission).

different from the phoria method. Moreover, the variations in the slopes that were possible in the nonlinear model were similar to the variation found experimentally. Thus, the nonlinear model, containing the deadspace operators DOF and PFA, could account for the discrepancies found between the two methods.

The FD curve (FD as a function of prism power, with eso and exo FD represented by + and −, respectively) has been used to clinically assess the integrity of binocular fixation, and generally has a curved shape[114] (Fig. 12A). The curved shape can be described as follows — as the prism stimulus increases from negative to positive (BI to BO) values, the FD curve begins at a relatively steep downward angle in the upper left quadrant, flattens out slightly and then crosses the zero stimulus point, and at the larger prism values resumes at a relatively steep downward angle in the lower right quadrant. Such a curve can be approximated by the transitions across the lines in the simulation

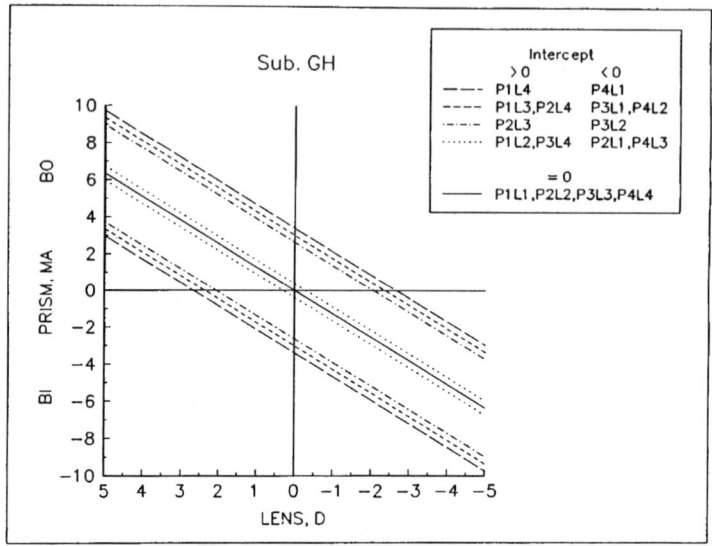

Fig. 15 Model simulation results obtained by inputing parameters for subject GH into the term, $B_{nonlinear}$, in Eq. 19. The form of the plot of prism versus lens is the same as that used by Ogle et al.[114] The legend indicates the combination of the difference between the prism and lens equations for a particular solution line, and whether the intercept is >, < or = to zero. For example, P1L4 represents the $B_{nonlinear}$ term obtained for the difference between VE_P under condition 1 of Table 3 and VE_L under condition 4 of Table 3, and its intercept is > 0. Note that there are fewer than 16 distinct lines in the graph because of the non-uniqueness of some of the solutions (reprinted from Hung,[65] p. 308, with permission).

plot (see Fig. 14). Indeed, such a curved shape is predicted by the multiple solutions obtained from operations on the two sides of the deadspace operator for both DOF and PFA. For one direction of the stimulus (e.g. BO prisms), the solution falls on one group of solution lines, but for the opposite direction of the stimulus (e.g. BI prisms), the solution falls on the opposite set of solution lines, with the transition between these solution lines providing the curved shape. A detailed discussion is provided by Hung[64] and involves the assumption of an increase in the size of PFA, and thus a wider separation between simulated lines, for a larger magnitude prism stimulus. Briefly stated, the curve begins at the uppermost prism line for the larger PFA in the upper left quadrant (see Fig. 14). It then moves to the uppermost line for the smaller PFA. After crossing the zero stimulus point, it moves across to the lowermost line for the smaller PFA at the lower right quadrant. Finally, at a large prism stimulus level, it moves to the lowermost line for the larger PFA. Thus, the nonlinear model is able to represent the complex phenomenon of an s-shaped FD curve.

The AC/A simulation results are also consistent with a previous modeling study on phorias.[64] For monocular viewing, the *vergence response* is called the disassociated phoria. On the other hand for binocular viewing, the *vergence stimulus* giving zero FD is called the associated phoria. The rationale of Ogle et al.[114] for using the associated phoria is that this VS value is assumed to indicate the point of oculomotor balance under binocular conditions. Also, they suggested that the associated phoria should be equal to the disassociated phoria. However, experimental studies showed that when the data are plotted as associated versus disassociated phoria, there is a broad, approximately Gaussian distribution rather than a clear direct relationship.[64] The modeling results showed that a linear model would illustrate a direct relationship between associated and disassociated phoria. However, the nonlinear model, containing the deadspace operators DOF and PFA, showed multiple solutions that would be consistent with the random broad distribution in the experimental associated phoria values.

Thus in general, the deadspace operators DOF and PFA in the nonlinear model are able to account both for the discrepancies between the associated and disassociated phoria, as well as between the AC/A ratio determined by the phoria and FD methods.

Proximal Model

Under naturalistic viewing conditions, as in our everyday surroundings, there is considerable information available with respect to target blur, disparity and proximity (as well as other binocular and monocular perceptual cues to depth) to drive the interactive accommodative and vergence systems.[16,39,58,60] Over the past three decades, the vast majority of experiments and models involving accommodation and vergence have considered target blur and disparity as the primary determinants of the overall response, with the static tonic inputs having a relatively small and constant contribution.[77,80,131,133,134] The proximal contribution has frequently been omitted, and where it did appear, it was often only described in a vague, qualitative manner such as an "injection" input term.[163]

However, over the past decade, there has been considerable interest in the role of target proximity, i.e. apparent target nearness, on oculomotor near

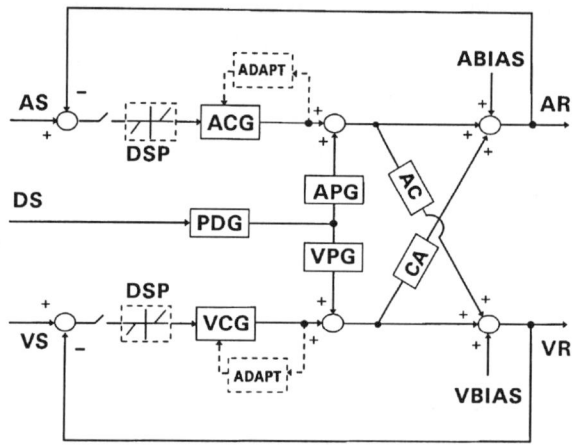

Fig. 16 Complete nonlinear static interactive dual-feedback model of the accommodative and vergence systems. For the *accommodative* system, the switch controls feedback to accommodation. With the switch open, the input to the accommodative deadspace operator (DSP, which represents the depth-of-focus) is zero. On the other hand, with the switch closed, the difference between accommodative stimulus (AS) and accommodative response (AR), or accommodative error (AE), is input to DSP. The output of DSP is multiplied with the accommodative controller gain, ACG, to give the accommodative controller output. The controller output is input to an adaptive element (ADAPT), which in turn controls the time constant of the accommodative controller. The distance stimulus (DS), or the distance of the target from the viewing subject, is input to the perceived distance gain (PDG) element, which represents the subjective apparent distance estimate. The PDG output then goes through the accommodative proximal gain (APG) element, which represents the contribution from target proximity. The outputs from ACG and APG are summed at the summing junction and are also cross-linked to the vergence system via gain AC. The accommodative bias (ABIAS), or tonic accommodation, is summed at the next summing junction along with the cross-link signal from the vergence controller output via CA. These four signals are added together to give the overall accommodative response, AR. Analogous descriptions of the parameters are applicable for the *vergence* system. The elements shown by dashed lines, ADAPT and DSP, are not used in the static model simulation (reprinted from Hung et al.,[74] p. 32, with permission).

responses. Recent objective evidence demonstrated that in the absence of visual feedback of target blur and disparity, proximity could drive the accommodative and vergence systems substantially. For example, the open-loop (OL) gain of both proximal accommodation (PA) and proximal vergence (PV)[139,140] has been found to be approximately 0.5, with target proximity being an effective stimulus from about 3 m to at least 20 cm.[132] Unfortunately, there has been a paucity of information with regard to its oculomotor effectiveness under natural viewing conditions, in which visual feedback regarding both target blur and disparity was readily available (i.e. closed-loop (CL) of both accommodation and vergence). Rosenfield et al.[137] have shown that steady-state accommodation to either a monocular far target (5 m) comprised of a blur-only stimulus (3 D), or a (normal) near target (33 cm) comprised of an equivalent dioptric stimulus mediated via blur-*plus*-proximity, resulted in statistically equivalent ARs. This was consistent with earlier results of other investigators,[1,110] and has recently been confirmed objectively by Shum et al.[153] and Jones.[88] In contrast, North et al.[112] found that in a CL cue conflict paradigm, PV (and PA) might influence the oculomotor response as much as the disparity input and more than the blur input.

Hence, a quantitative analysis of a comprehensive proximal model of accommodation and vergence was performed to determine the contribution of proximity to the AR and VR under OL and CL conditions[74] (Fig. 16). Equations were derived, and after some manipulations, AR and VR were solved as a function of the various parameters under different combinations

Table 4 Proximal Model Parameter Values.
(Adapted from Hung et al.,[74] p. 34, with permission.)

Perceived distance gain	PDG	0.212
Accommodative proximal gain	APG	2.100
Vergence proximal gain	VPG	0.067
Accommodative controller gain	ACG	10.0
Vergence controller gain	VCG	150.0
Accommodative convergence	AC	0.80 MA/D
Convergence accommodation	CA	0.37 D/MA
Tonic accommodation	ABIAS	0.61 D
Tonic vergence	VBIAS	0.29 MA

of OL and CL conditions. Parameter values were obtained based on experimental data (Table 4). It was found that the relative contribution of PA to the overall AR under accommodation CL (A_{CL}) and vergence OL (V_{OL}) was given by:

$$\left[\frac{PA\ term}{overall\ AR}\right]_{A_{CL}, V_{OL}} = \frac{0.0409 * DS}{0.950 * DS + 0.0555} \quad (22)$$

Also, the relative contribution of PA to the overall AR under accommodation and vergence CL was given by:

$$\left[\frac{PA\ term}{overall\ AR}\right]_{A_{CL}, V_{CL}}$$

$$= \frac{[APG*(1+VCG) - APG*VCG*AC*CA + VPG*CA]*PDG*DS}{\substack{[ACG*(1+VCG) - ACG*VCG*AC*CA]*AS + VCG*CA*VS \\ +[APG*(1+VCG) - APG*VCG*AC*CA + VPG*CA]*PDG*DS \\ +(1+VCG)*ABIAS - VCG*CA*VBIAS}}.$$

(23)

Similar expressions were obtained for the vergence contribution under different conditions. Using parameter values from Table 4, the relative contributions of PA and PV to the overall static AR and VR are shown in Fig. 17.

The results of the present experimentally-based model simulation investigation clearly demonstrated that when both retinal defocus and disparity feedback were rendered ineffective, i.e. OL, the proximal motor contribution was quite large (~40 to 90%). Thus under these conditions, the oculomotor response was primarily based upon the higher-order, non-retinal, perceptual attributes of the stimulus.

However, when only one of the systems was rendered ineffective, the relative proximal contribution to that system was much smaller. It was less than 5% for accommodation and 7% for vergence under normative conditions. An exception was the A_{OL}, V_{CL} condition, where the relative contribution of PA was fairly large (~25 to 40%). The magnitude of the relative PA contribution under this condition reflected the rather high value of accommodative proximal

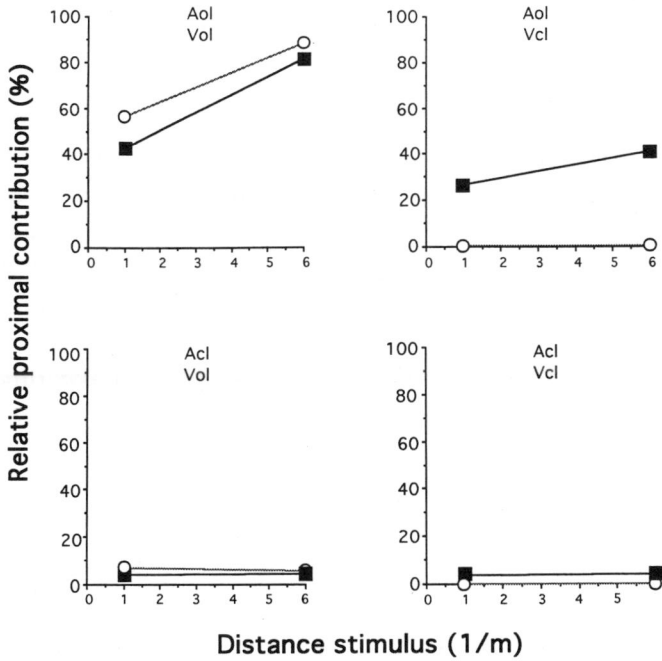

Fig. 17 Relative contribution of the proximal accommodative terms to the overall accommodative response (closed squares) and the proximal vergence terms to the overall vergence response (open circles) for distance stimuli of 1 and 6 (1/m) under different accommodative and vergence loop conditions (reprinted from Hung et al.,[74] p. 37, with permission).

gain (APG = 2.10), which was able to drive the accommodative output directly when the accommodative feedback loop was rendered ineffective. Thus, the vergence-driven AR in the absence of blur information appeared to provide a substantial proximal contribution. In addition, sensitivity analysis for all conditions using either higher or lower gain values showed relatively little effect on the proximal contribution. It never exceeded 9% for accommodation and 12% for vergence.

Moreover, when visual feedback to both retinal defocus and disparity were present, these inputs dominated the vergence and accommodation oculomotor response. The proximal motor contribution under this condition was thus very small for accommodation (~4%) and was negligible for vergence (~0.04%).

These relative inputs were consistent with the normal OL gains of the respective systems. Accommodation, with an ACG of only 10, had allowance for relatively more proximal drive than disparity vergence, with its very high gain (VCG) of 150. Sensitivity analysis revealed that either a large increase or decrease in system gain had relatively minor impact on percent proximal output. It never exceeded 10% for accommodation and 1% for vergence. Thus, under typical naturalistic viewing conditions, proximal information played only a small role in determining the overall steady-state AR and VR levels, with motor responsivity now being primarily based on the lower-order, retinal, physical attributes of the stimulus.

The present model results, which were derived from experimental data, demonstrated that under normal binocular viewing conditions, the effect of target proximity on static oculomotor responsivity was minimal. This finding was consistent with and predicted by the narrow limits imposed on the steady-state oculomotor response by the deadspace operators, namely the DOF for accommodation and PFA for vergence.[70] Since control aspects for both of these deadspace operators were based entirely on visual feedback, the influence of target proximity as well as any other source of target information related to its perceived distance must have minimal impact on the final steady-state portion of the oculomotor response, otherwise diplopia and/or blur would ensue. This, however, contrasts with the view of a recently proposed model of the interactive accommodative and vergence oculomotor systems.[146] Their model suggested that under normal binocular CL viewing conditions, the effects of target proximity would be small when the stimulus change was small, and large (even dominant) when the stimulus was large. However, the former was based on simulations that exhibited unrealistic transient oscillations, whereas the latter was based only on conjecture. Also, their model could not account for experimental results that showed normal AR[11,72] and VR[76,82] when proximal cues were either held constant or were totally absent in the stimulus array.

Proximity probably has many sources of input which give rise to the overall visual perceptual sensation of "apparent distance/nearness of a target." In addition to the monocular (overlap, texture gradient, size, etc.) and binocular (disparity innervation/efference copy) visually-related information,[84] other inputs may include kinesthetic/tactile information such as when we touch or grasp an object,[126] information from other senses such as audition and olfaction,[126] memory of object distances and propinquity (as in total or near total darkness),[130] and accommodative innervation/efference copy.[42]

Under typical everyday viewing conditions, blur and disparity play a dominant role in the generation of the steady-state motor response in visually-normal individuals. Other cues, as described above, also play a role by assisting the visual system in target localization and acquisition to provide a harmonious and coordinated response. Indeed, visual and non-visual cues act in concert to reinforce and confirm the dominant retinal blur and disparity information. One of these cues, proximity, may contribute in at least two possible ways. First, as indicated by the simulation results,[74] proximally-driven responses might be relatively large and effective under either monocular or unfused conditions with poor quality AS present, such as might occur with low contrast/ low spatial frequency objects, eccentric retinal targets, and/or mesopic luminance operating levels.[16] And second, as a general rule in visual perception, one's response to a new target is more stable, effective, and accurate if all information is harmonious in nature and thus the various cues, including proximal factors, positively reinforce one another.[60,84] For example, as an object is moved towards an observer, monocular and binocular information regarding this change, including blur, binocular disparity, size, texture gradient, overlap, distance or proximity, etc., as well as efference copy, and proprioceptive and kinesthetic information, all change in a consistent and predicted manner.

Conversely, when any information related to a target change is conflicting and/or contradictory, so that there is "cue disharmony," the response might be adversely affected and altered.[112] For example, North *et al.*[112] used a non-naturalistic, cue-conflict stimulus paradigm in which the relative contributions of blur, disparity and proximity were estimated. Under these unusual conditions, proximity (45%) and disparity (41%) contributed much more to the overall VR than blur (14%). However, it should also be noted that all of the measurements were obtained either with the vergence system OL or in the absence of central field visual feedback, with these conditions biasing in favor of proximity and against blur.[16]

Given the potential role of proximity under specific conditions in the normal environment, how might this carry over to the clinical environment? Proximity may again play an indirect but important role. In various clinic populations (such as strabismics and others with binocular vision dysfunctions) and clinic test environments (such as in the phoropter or the synoptophore, which are special instruments for measuring static accommodation and vergence), proximity may have specific and directly measurable impacts[60,168] that are predicted by our model. For example, due to the awareness of nearness in the synoptophore test environment, patients typically exhibit more relative esophoria or tropia in such "instrument space" than in "free space" in this OL vergence condition.[60] And in patients with low oculomotor system gain, i.e. convergence and/or accommodative insufficiency, the proximal motor component has the potential to be relatively larger and have greater influence,[16] which is consistent with our sensitivity analysis results. Finally, in such patients undergoing optometric vision therapy, our results suggest that proximity, along with the other cues, may provide valuable information to guide the motor response as well as to assist in binocular spatial localization, for example during fusion training with stereo vectograms, until both accommodative and vergence responsivity normalize, and thus cue harmony is re-established.[28,60,168]

Sensitivity Analysis of Accommodation and Vergence Interactions

To investigate the effect of parameter variation on AR and VR, a sensitivity analysis was performed on the interactive dual–feedback model[70] (Fig. 13). For each parameter listed in Table 2, two simulated stimulus paradigms were performed. First, the lens value was varied between ± 2.5 D while the prism value was held at 2.5 MA, and the accommodative error (AE) was calculated. Second, the prism value was varied between 25 BI and 25 BO while the lens value was held at 2.5 D, and the VE (or FD) was calculated. Moreover, this was repeated at 50% and 150% of the nominal value, while all other parameters were held constant at their nominal values.

The results indicated that an element *within* a feedback loop will tend to have a relatively modest influence on overall system sensitivity due to feedback regulation of its output. This is in contrast with an element that *links* two feedback loops and therefore has multiplicative influences that enhance the responses of both loops, thus having a greater impact on overall system sensitivity.

Indeed, it was found that the model was most sensitive to variations in the crosslink gains AC and CA. These gain elements interconnect the two feedback systems and thus determine the extent of mutual interaction between accommodation and vergence. Since the output from each controller is multiplied with its crosslink gain element to serve as an input to the fellow system (Figs. 11A and B), it effectively increases these interactive influences. That is, the multiplicative effects of each parameter, both within and across each motor system, contribute to and therefore influence the overall response to achieve steady-state system stability. It can be shown[70] that the main effect of increased AC or CA is to increase the AE and VE, with this occurring to a much greater extent than from variation in the other parameters. Another way to look at this influence is to inspect more closely the equation for VE under CL accommodation and vergence (see Eq. 14). If the denominator of Eq. 14 is set to zero, we have

$$(1+ACG) \bullet (1+VCG) - ACG \bullet VCG \bullet AC \bullet CA = 0. \qquad (24)$$

Dividing both sides of Eq. 24 by $(1+ACG) \bullet (1+VCG)$ gives

$$1 - \frac{ACG}{1+ACG} \bullet AC \bullet \frac{VCG}{1+VCG} \bullet CA = 0. \qquad (25)$$

It can be shown that this is equivalent to[78] (see Eq. 13):

$$1 - AC/A \bullet CA/C = 0 \qquad (26)$$

where AC/A and CA/C are the accommodative convergence to accommodative stimulus ratio and the convergence accommodation to convergence stimulus ratio, respectively. This indicates that if AC/A $\bullet$ CA/C = 1, the denominator in Eq. 14 would go to zero, and VE would rapidly grow to an infinite value. Such a large VE would correspond to the oculomotor disorder of strabismus ("cross-eyed" or "wall-eyed").[49] Indeed, examination of the literature where both AC/A and CA/C were measured in subjects with normal binocular vision[40,77] shows that the product AC/A $\bullet$ CA/C range in value from 0.21 to 0.68 (see Table 2), which are below the value of 1.0. On the other hand, this also suggests that individuals with AC/A $\bullet$ CA/C product close to 1.0 are likely to develop vergence anomalies. However, if either or both AC/A and CA/C could be modified early in life, before anomalous correspondence developed, through optical or surgical means to bring the product to below 1.0, the anomaly may be either reduced or eliminated. Thus, this finding has important clinical implications for the amelioration of such conditions as strabismus and vergence dysfunction.

The model was only moderately sensitive to variations in the controller gains ACG and VCG, as well as the tonic terms ABIAS and VBIAS. With respect to controller gain, the results can be explained by the fact that for an isolated feedback control system, the overall CL gain attributed to the controller is of the form G/(1+G) (see Eq. 1B). Such a CL term is inherently only moderately sensitivity to changes in G. Indeed, it is a basic feature of such a system to maintain stability despite larger fluctuations in the controller gain, for example, as may occur in neurological disease.[24,98] In addition, since each controller primarily governs its own feedback loop and has only an indirect influence on the fellow loop, changes in G would have a relatively small influence on the fellow system response. Similarly, it can be shown that for an isolated feedback control system, the tonic term has only a relatively minor effect on the CL response,[77] having the form of 1/(1+G) * ABIAS. Thus, it is not surprising that the model is also only moderately sensitive to changes in the tonic terms.

On the other hand, the model was quite insensitive to variations in the deadspace elements, and this is in good agreement with earlier experimental results.[114,128] Perhaps the system uses feedback to adjust AE and VE to maintain a relatively constant deadspace element output despite changes in the deadspace limits. This lack of sensitivity is also seen in the robustness of the static physiological responses. For example, the accommodation system exhibits relatively small response changes with reasonably large increases in the DOF as it apparently attempts to maintain the output of the accommodative deadspace element approximately constant, until the pupil diameter is reduced to below 1 mm.[128] Similarly, for the vergence system, its error is maintained relatively constant despite large (about ± 10 prism diopters) changes in the VS, with associated changes remaining essentially within Panum's fusional area, at least in asymptomatic visually normal individuals.[114]

How might these sensitivity analysis results provide insight into the mechanisms producing abnormal binocular vision, such as strabismus and especially esotropia, which is found in 3% of the generally population[61] and up to 50% in special populations such as cerebral palsy.[34] One can regard the DSP term for both accommodation (DOF) and vergence (PFA) as a "tolerance for error." Therefore if DSP increases, the allowable error should increase to some extent. With regards to the vergence system, consider a strabismic individual with an overall, combined central foveal/diplopia point binocular suppression scotoma.[21] If small central disparity stimuli are presented along the midline, a symmetric disparity VR is not elicited, as the stimulus in the strabismic or deviated eye falls within the scotoma.[89] Therefore, for all practical purposes, it is suppressed and is not processed as a binocular retinal disparity. This results in the absence of a disparity VR. However, if a very large field (~50 deg) midline disparity stimulus is now presented to the same subject, with the disparity stimulus now extending beyond the scotoma and onto the normal peripheral retina having large PFA than centrally, a symmetric disparity VR now occurs.[6] The response is reduced by only 30% as compared with that of a visually normal individual. Thus, the 300% (or more) larger PFA in the retinal periphery[114] has only a moderate impact (i.e. 30%) on the overall disparity VR amplitude. With regards to the accommodation system, both amblyopes[21] and nystagmats[118] have a twofold increase in DOF as compared with normal individuals, and yet this produces only a modest effect on the slope of the accommodative stimulus-response function and overall accommodative accuracy. For both vergence and accommodation, the increase in DSP, while appearing to be quite large (e.g. two times of 200% greater), in actuality only represents a relatively small portion (<15%) of the typical amount of stimulus change (e.g. a few diopters or meter angles) to either system. Thus the simulation sensitivity results are consistent with physiological findings.

With regards to the crosslink terms which appear to be most sensitive to change, the results may provide new insight into the development of accommodative esotropia in which a high AC/A ratio is the hallmark sign. Relatively small changes in the parameter AC results in vergence overdrive with consequent large changes in vergence steady-state error. Inability to compensate habitually via fusional divergence could result in the development of esotropia at near distances.

With regards to the controller gain terms, the simulations showed that they have only moderate impact on system responsivity. For vergence, patients with symptoms related to near work exhibited only moderate gain reduction in disparity vergence as compared with their asymptomatic counterparts.[75] For accommodation, amblyopes have demonstrated markedly reduced gain, but with only moderate effect on the slope of the AS/R function.[17] Thus, in patient studies in which this parameter was indeed abnormal, only moderate effects on motor responsivity were found, as predicted by the simulation results.

Lastly, with regards to the tonic terms, the model equations clearly reveal that this parameter has only a moderate impact on system response amplitude,[77] especially at near distances, where most patient symptoms are manifested. This was indeed found in clinical studies on accommodation and vergence.[17,75]

Dynamic Analysis Techniques

Main Sequence

Dynamic oculomotor responses exhibit a characteristic monotonically increasing relationship between amplitude and peak velocity called the main sequence. The normal average main sequence slope in (deg/sec)/deg is ~2.5 for accommodation,[72] ~4.0 for vergence,[83] and ~50 for saccade.[3,107] Thus, for example, the peak velocity for a 4 deg saccade is about 200 deg/sec, but that for a 4 deg vergence is only about 16 deg/sec. Data falling within the main sequence are considered normal, whereas those falling outside the main sequence are considered abnormal. Therefore, the main sequence provides a useful indicator of the integrity of the accommodative neural controller signals (it is assumed that the plant is normal and non-varying due to age or disease).

Accommodation System — Root Locus Analysis

Campbell et al.[10] found a larger peak at 2 Hz and a smaller peak at 0.5 Hz in the accommodative frequency spectrum. To determine whether this was caused by system instability oscillations, a root locus analysis was applied to the accommodation system. The root locus analysis technique[31] is a general method for describing the stability of a feedback system and the variation in its stability as a function of the forward-loop gain. This method is well suited for evaluating the stability of the accommodation system and its dependence on ACG. However, the root locus method is only applicable to linear systems; hence, for this analysis a linear approximation is used in place of the nonlinear deadspace operator, or the DOF. As the accommodation system generally operates on one side of the DOF (the low output side for normal near viewing, leading to the descriptive term "lag of accommodation"),[110] the deadspace operator can be appropriately replaced by a linear bias or "offset" term[77] (Fig. 11B).

A root locus computer program developed by Krall and Fornaro[91] was used to analyze the dynamic stability characteristics of the accommodation model.[81] The unique feature of this root locus program is that it allows for a delay element (to simulate latency) in the feedback loop. The program plots the location of the closed-loop poles (equal to the roots of the denominator of the transfer function, so that when the system operates near the root value, the transfer function output rapidly grows to an infinite value, hence the name pole) as a function of gain K for forward-loop transfer function of the form, $Ke^{-\tau s} G(s)$, or overall closed-loop transfer function of the form

$$\frac{Ke^{-\tau s}G(s)}{1+K^{e-\tau s}G(s)}. \tag{27}$$

For the accommodative model analysis, the following parameter values were used: controller time constant equal to 6 sec;[93,122] plant time constant of 300 msec;[162] and a total time delay of 350 msec.[169] Thus,

$$Ke^{-\tau s}G(s) = \frac{Ke^{-0.35s}}{(1+6s)(1+0.3s)}. \tag{28}$$

The root locus plot of the linearized accommodation system with forward-loop transfer function given by Eq. 28 is shown in Fig. 18, where the horizontal and vertical axes are the σ and $j\omega$ axes, respectively. From this plot, we note that the gain corresponding to the root locus at the $j\omega$ axis is equal to 21, which gives the maximum gain before instability oscillation occurs (i.e. poles in the right half of the s-plane, or to the right of the vertical $j\omega$ axis, corresponding to a time domain function which grows rapidly, and would mean the system is unstable). The predicted frequency for instability oscillation of 0.45 Hz (or 2.8 rad/sec) is near the lower accommodative spectral peak at 0.5 Hz, but far below the higher peak at 2 Hz.[10] This indicates that loop instability is not a source of higher frequency accommodative oscillations. However, the smaller peak at 0.5 Hz may in fact correspond to the closed-loop oscillation frequency of 0.45 Hz indicated in the root locus plot. Moreover recently, Winn et al.[170] found a significant correlation between the arterial pulse frequency and the accommodative higher frequency peak. Thus, the lower

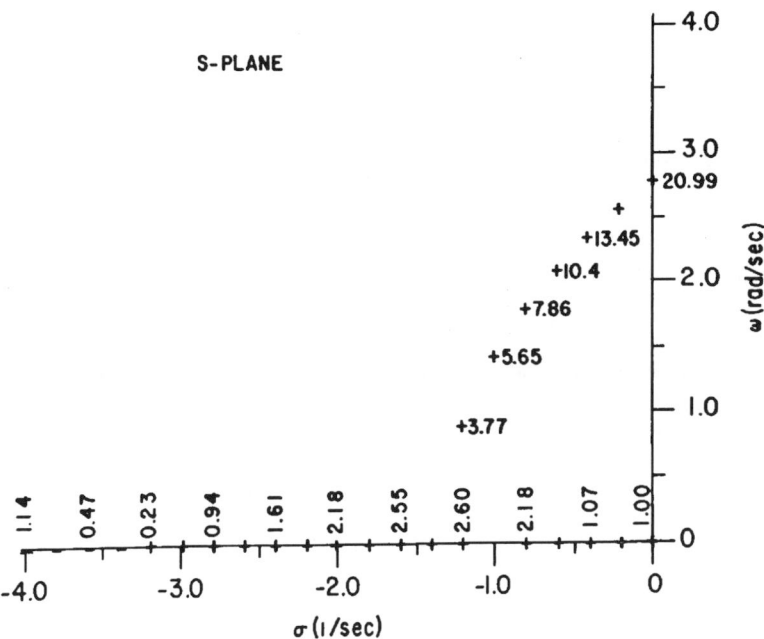

Fig. 18 Root locus plot showing stability characteristics of accommodation. Closed-loop pole locations are shown for overall closed-loop transfer function of Eq. 24 with the value of gain K displayled next to the corresponding root locus pole. Positive gain is marked with plus symbol along with numerical value, and negative gain is marked with minus symbol. Poles lying within region near the left horizontal axis indicate a damped closed-loop step response, and poles at the vertical axis describe an unstable system (reprinted from Hung,[81] p. 597, with permission).

accommodative frequency peak is most likely associated with neurologically controlled feedback instability oscillations, whereas the high frequency peak is an epiphenomenon due to the effect of arterial pulse on lens motion that is picked up by the recording optometer.

Vergence Dual-Mode Dynamic Model

As described previously, the vergence system responds to a target change in depth by oppositely directed rotations of the two eyes.[125,147] It exhibits a latency of about 180 msec for convergence and 200 msec for divergence.[149]

Its dynamics are relatively slow compared to the saccadic eye movement system,[3] and has a time constant of about 180 msec for convergence and 250 msec for divergence.[83] It is this relatively slow dynamics, along with its relatively long latency, that have made it difficult to model vergence as a continuous feedback system. Indeed, when open-loop parameters based on experimental data were input into a continuous feedback model, it exhibited instability oscillations.[80]

Additional evidence that a continuous feedback model is inappropriate can be found in the experimental vergence responses to ramp stimuli.[148] For ramp velocities less than 2 deg/sec, the vergence system tracked with a smooth ramp response. On the other hand, for ramp velocities greater than 9 deg/sec, the vergence system tracked with step-ramp and multiple-step responses. The steps in the step-ramp responses were not simply continuous delayed movements reflecting an earlier disparity error, but were accurate step movements that matched the instantaneous ramp stimulus amplitude at the time corresponding to the end of the initial vergence movement. These vergence results suggested a preprogramming process in which the amplitude of the initial step of the step-ramp response was proportional to the target velocity. At the end of the step movement, a slow continuous feedback controller took over to adjust the vergence response to be within a small acceptable error limit. The preprogrammed process operated under an open-loop condition (i.e. the output was directly driven by the input without any feedback), which precluded feedback interactions involving the latency component. This allowed for a stable step movement in an otherwise unstable feedback system. The slow controller operated when the disparity, or vergence error, was small, so that a relatively small slow controller gain could maintain the vergence response. The overall result of this two-stage process over the entire response amplitude was a fast and stable dynamic response along with an effectively high gain steady-state response.

Thus, to maintain accuracy as well as stability, a dual-mode model for the vergence eye movement system was developed and simulated using MATLAB/ SIMULINK (Fig. 19). It consisted of a fast open-loop component and a slow closed-loop component. The fast component accounts for most of the vergence response, while the slow component uses feedback to reduce the small residual disparity to a minimum. The sum of their outputs provides the command signal. The fast component (see FAST block, Fig. 19; and flow chart, Fig. 20A) is

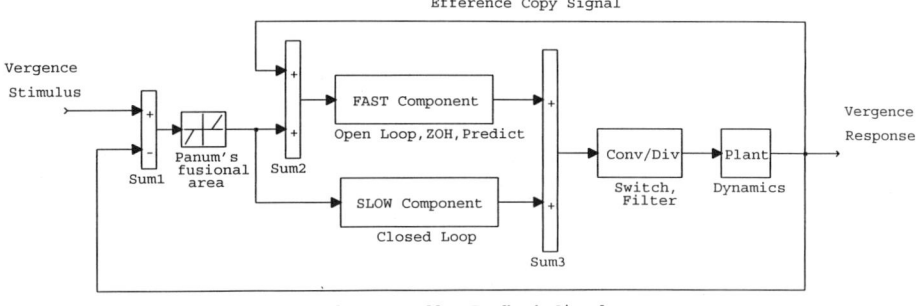

Fig. 19 Block diagram of the vergence system used in the MATLAB/SIMULINK simulations. The difference between vergence stimulus and response, or vergence error, is input to a deadspace element, which represents Panum's fusional area. The output of the deadspace element is summed with the efference copy signal, resulting in a signal equal to the actual stimulus. This signal is used to drive the fast component. The fast component operates in an open-loop manner, uses a zero-order hold (ZOH), and has a predictive capability for periodic stimuli. The fast component output is input to the convergence/divergence switch and filter, which in turn drives the oculomotor plant. The output of the plant provides the efference copy signal, which takes into account the effect of the filter and plant. The output of the deadspace element also drives the slow component. The slow component operates over a smaller range of vergence error amplitudes and velocities. The output of the slow component is input to the convergence/divergence switch and filter, which in turn drives the oculomotor plant. The output of the plant is fed back and is substracted from the vergence stimulus to provide the error signal to the deadspace element. Thus, the slow component operates under a closed-loop condition. The fast and slow components operate over different stimulus regimes so that when one is active, the other is disabled. This provides robustness in the model response (reprinted from Hung,[62] p. 61, with permission).

driven by the sum of the error signal (equal to the output of the disparity signal through the deadspace operator, with limits = ± 2.5 min of arc,[70] which represents PFA[121] and the efference copy[35,173] signal from the fast component output. The efference copy is an internal neurological signal that is associated and coordinated with the intended motor movement, so that the expected motor response can be part of the overall motoneuronal control program. Thus, since the fast component output is an open-loop movement that approximately equals the stimulus amplitude, the efference copy signal notifies the neurological control program of the slow component to respond only to the residual error. The fast component open-loop drive is important because it maintains stability

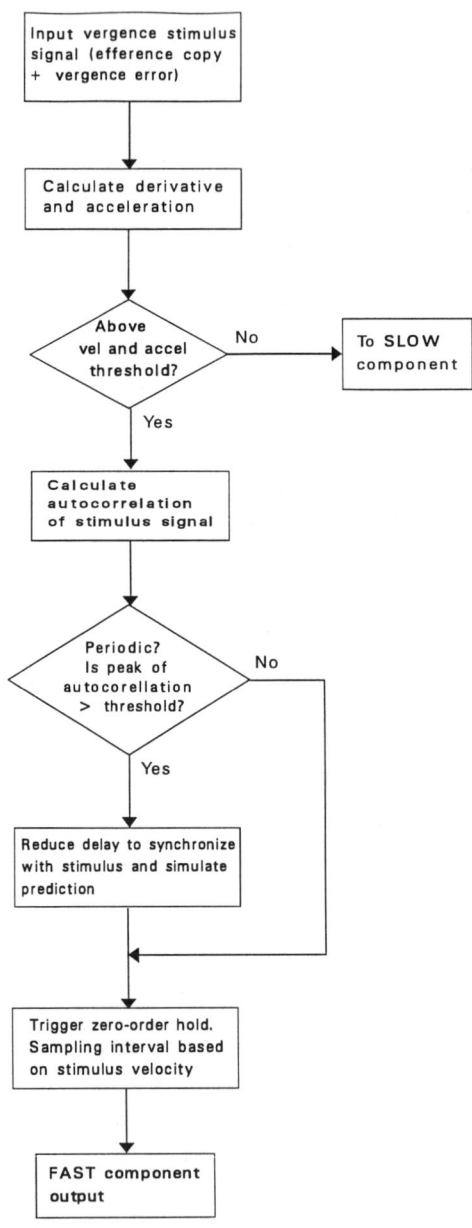

Fig. 20A Flow charts for the FAST elements of the model shown in Fig. 19 (reprinted from Hung,[62] p. 62, with permission).

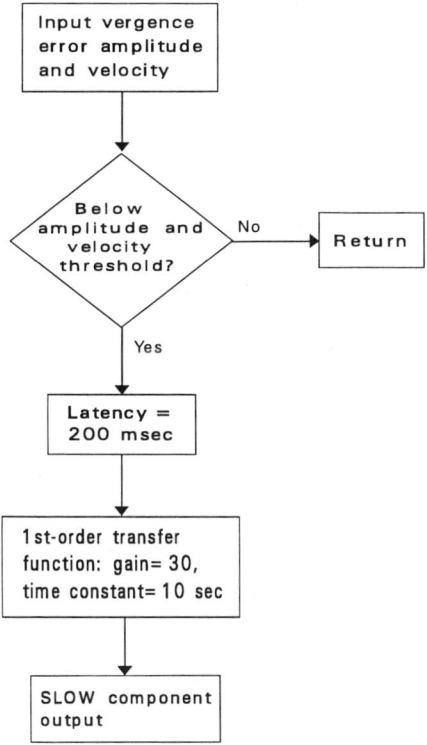

Fig. 20B Flow charts for the SLOW elements of the model shown in Fig. 19 (reprinted from Hung,[62] p. 62, with permission).

in the presence of a relatively long latency (~ 200 msec) and the requirement of an accurate initial step response. Such accuracy corresponds to very high gain in a feedback control system, which would have otherwise resulted in instability oscillations. The fast component open-loop response therefore accounts for most of the step response amplitude, with the remainder being taken up by the closed-loop slow component. The slow component (see SLOW, Fig. 19; and flow chart, Fig. 20B) operates over smaller amplitude and velocity ranges and uses negative feedback to provide the error signal for the controller. Significantly, the FAST and SLOW components operate under separate stimulus regimes, so that when one is active the other component is not. This provides robustness of the vergence response, since there is no cross

Table 5 Selected Vergence Model Parameters
(Adapted from Hung,[62] p. 63.)

Parameter	Value	Description
Athrsh	2500	Acceleration threshold (deg/s)/deg for trigger of fast component for pulse and step.
Pcross	5 or 20	Percentage of period for maintenance of periodicity of cross-correlation between stimulus and response, for low and high velocities, respectively.
Pthrsh	0.75	Amplitude threshold (deg) for trigger of fast component for ramp and sine.
Samptime	0.2 to 0.5	Sampling interval (sec) of ZOH for high to low stimulus velocities, respectively.
Tflag	0 or 1	Global flag for trigger of fast component and disable of slow component, and *vice versa*.
Thrsh1	0.25 or 0.45	Threshold for initial detection of peak of autocorrelation for low and high velocities, respectively.
Thrsh2	0.15 or 0.40	Threshold for detection of maintained peak of autocorrelation for low and high velocities, respectively.
Vthrsh	2.50	Velocity threshold (deg/s) for trigger of fast component for ramp and sine.

interference (other than the residual decay of the slow component when it is disabled) between the fast and slow component outputs. Indeed, in test simulations (not shown) where both fast and slow components are simultaneously active, the model exhibits unstable, non-robust, and sometimes bizarre responses. The outputs of the fast and slow components are summed to provide the input to a convergence/divergence switch and filter. The switching is based on the sign of the velocity. The output of the switch and filter provides the command signal that drives the oculomotor plant. The following describes in detail the Fast.m function in the FAST block and Slow.m function in the SLOW block (Fig. 19).

The fast component function, Fast.m, consists of a zero-order hold (ZOH; hold value for one sampling interval) element which responds to changing stimuli, such as ramps that are above the position (Pthrsh) and the velocity

(Vthrsh) thresholds, or steps above the acceleration (Athrsh) threshold (see Table 5 and Fig. 20A). The sampling interval (Samptime) varies ranging from 0.2 to 0.5 sec for high and low velocity ramp stimuli, respectively. A sudden subsequent rapid change in the stimulus, however, will re-trigger the ZOH. This provides model responses that are consistent with experimental ramp and square-wave responses. For the pulse stimulus, a double change in sign of acceleration (to distinguish it from the single change in sign for a step) within a 200 msec interval will also trigger a downward movement (thus overriding the ZOH) and result in a delayed (200 msec) pulse response that follows the time course of the pulse stimulus. This provides model responses that are consistent with the experimentally observed increase in response pulse amplitude and duration with increased stimulus pulse duration. This behavior had led some earlier investigators to erroneously conclude that the vergence system exhibited continuous feedback control for all stimuli. For a sinusoidal stimulus, the fast component operates in conjunction with the slow component to provide the sinusoidal response. Moreover, the latency, which decreases from a maximum of 0.5 sec for slow ramps to a minimum of 0.2 sec for pulse, step and fast ramps, is embedded in the Fast.m function.

Also within the function, Fast.m, the autocorrelation coefficients of the stimulus time course is calculated at each time step. The autocorrelation provides a means to quantify the periodicity in a waveform[158] to allow for prediction to drive the response. If a positive peak value (away from the zero shift position) is detected (correlation coefficient > Thrsh1), this is considered to have detected a periodic stimulus, such as a sinusoid. The periodic interval is equal to the time difference from zero shift to the shift at the positive peak. The stored periodic stimulus pattern, with a time shift advancement simulating prediction, is then used to drive the ZOH element of the fast component. The maintenance of periodicity is checked in two ways. First, the autocorrelation peak (away from the zero shift position) of the estimated stimulus timecourse is compared against a threshold (Thrsh2, which is slightly lower than that for detection). Second, the cross-correlation between the stimulus and response is calculated once every period to determine whether there is a large phase shift (i.e. whether the central peak is deviated from the center by a threshold amount by Pcross percent of the period). If the threshold is exceeded in either of the above conditions, the periodicity is considered to have been lost. If such a subsequent loss of periodicity is detected, the operation mode is reverted back to normal and is ready to detect any subsequent periodicity.

The slow component function, Slow.m, responds to changing stimuli below the position (Pthrsh) and velocity (Vthrsh) thresholds (see Table 5 and Fig. 20B). These include slow ramps, low frequency sinusoids, and the level position at the end of step responses. The slow component latency = 0.20 sec, gain = 30, and 1st-order element time constant = 10 sec.[94] The slow component uses negative feedback to fine-tune the response. This simulates the experimentally observed fusional process.[125] The fast and slow component functions, Fast.m and Slow.m, communicate via the global parameter Tflag. When Tflag = 1, the fast component is active and the slow component is disabled. On the other hand, when Tflag = 0, the opposite condition holds.

Thus, the fast and slow components are placed in parallel in the forward loop, with each operating over a range of stimulus amplitudes and velocities. The fast and slow component outputs are summed to provide the signal to the convergence/divergence switch and 1st-order filter (time constant = 0.07 sec for convergence and 0.15 sec to divergence; note that these elements are input to the plant to provide the overall dynamics) whose output provides the command signal that drives the oculomotor plant.[129] The plant output serves as the efference copy signal which is fed back and summed with the error signal to provide the input to the fast component. The signal from the plant, rather than the fast component output, is used because it provides a more accurate value for the fast component output. This is appropriate since only one component (fast or slow) is active at a time, and little or no inappropriate slow component residual signal is present in the efference copy signal. For the slow component, its output is fed back as a negative feedback signal and combined with the stimulus to result in the error signal that serves as the input to the slow component.

Simulations were performed on an 486-PC with 8 MB RAM operating at 66 MHz, Windows 3.10, MATLAB 4.2c, and SIMULINK 1.3c. Stimulus functions that were not available in Blocklib.m of SIMULINK, such as pulses, ramp-stop, and square-wave (above zero level), were written as *m*-files and called by *s*-function blocks.

The model simulation results exhibited behavior that was similar to that found experimentally.[80,83,92,148,175] For example, the ramp responses showed

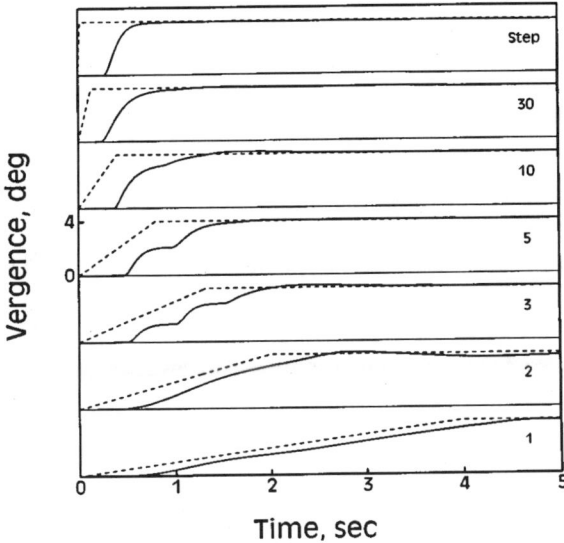

Fig. 21 Model responses to positive (convergent) ramp stimuli (velocity in deg/sec shown at right of traces) with maximum amplitude of 4 deg. Dotted line = stimulus, solid line = response (reprinted from Hung,[62] p. 65, with permission).

transitions from smooth tracking by the slow component for slower ramps (1 and 2 deg/sec), to staircase-steps by the fast component for intermediate velocity ramps (3, 5 and 10 deg/sec), and single-steps by the fast component for higher velocity ramp and step stimuli (Fig. 21). This held true for both positive and negative ramps, with convergence exhibiting faster dynamics than divergence responses. Note also that there was a gradual decrease in latency with increasing ramp velocity, which was consistent with experimental results.[80,148] Also, sine-wave responses showed a transition from smooth tracking by the slow component for lower frequency stimuli (0.1 and 0.2 Hz) to combined fast and slow component responses for higher frequency stimuli (above 0.2 Hz) (Fig. 22). There was a transition from a non-predictive mode for the lower frequency stimuli to a predictive mode for the higher frequency stimuli. After about two cycles, the predictive mode came into effect and there was a reduction in latency in the sine-wave responses.

There were several important assumptions in the model. The fast component was assumed to behave as a ZOH element. This provided the

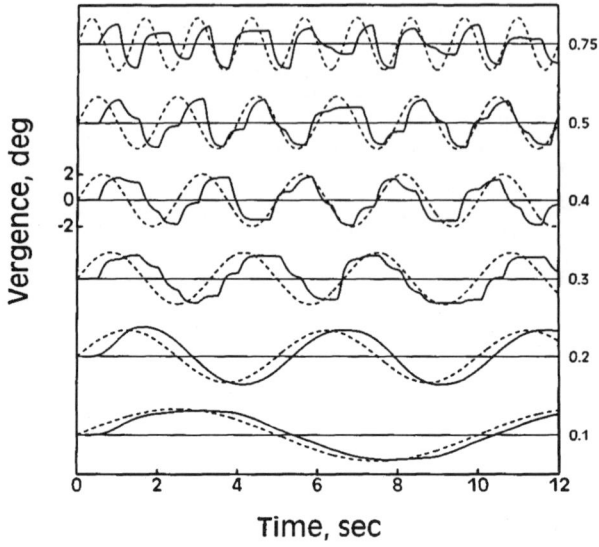

Fig. 22 Model responses to sine-wave stimulation (frequency in Hz shown at right of traces) for ± 2 deg peak-to-peak amplitude. Dotted line = stimulus, solid line = response (reprinted from Hung,[62] p. 66, with permission).

multiple-step responses to intermediate velocity ramps seen experimentally. Each of these steps have been shown to fall on the main sequence for vergence steps.[83,148] Without such an element, an oscillatory (i.e. sinewave-like) ramp response would not resemble the experimental data, and significantly, individual movements would not fall on the main sequence. Although Rashbass and Westheimer[125] termed their responses to ramp stimuli "oscillations," these have been shown to be multiple-step responses.[148] Another assumption was that the efference copy signal contained the plant dynamics following the output of the fast component. Such a process may occur in the brain by means of a neural network trained to represent the overall dynamics of the eye movement.[95] This would provide an appropriate representation of the eye movement response, which when summed with the disparity signal, gave an accurate representation of the stimulus. The third major assumption was that when the fast component was active, the slow component was disabled and *vice versa*. This distinct temporal separation of fast and slow component control was critical in that it permitted reasonable and robust estimation of the parameters. Otherwise, if both fast and slow component were simultaneously active, with one being open-loop and the other closed-loop, unpredicted and unusual behavior often occur, with an inability to control system performance as parameters were varied. The separation of the fast and slow component was consistent with neurophysiological results in monkeys showing near and far cells that respond maximally to targets 1 deg or more either in front of or behind the fixation plane, and tuned excitatory and inhibitory cells that respond maximally to target near the fixation plane.[123] Thus, the conditions for stimulating the fast and slow components corresponded to those for driving the far and near cells and the tuned excitatory and inhibitory cells, respectively. Finally, it was assumed that the peak (away from the zero shift position) of the autocorrelation function for the estimated stimulus time course could be used to assess periodicity.[158] This signal processing procedure was reasonable since the peak was relatively independent of the shape of the waveform but was dependent on its periodicity. Also, the height of the peak gave a measure of the accuracy of the estimated period.

The saccadic eye movement system has been shown to exhibit sampled-data behavior,[158] and the pursuit system has been shown to exhibit prediction.[90,158] Also, the accommodative system has been shown to exhibit sampling properties.[76] As indicated above, since the Fast.m function for the vergence system contained sampling and prediction, it could be modified to reflect the characteristics of the versional and accommodation systems. Therefore, the dual-mode model was useful not only for simulating vergence dynamics, but could be extended to other oculomotor systems as well.

Accommodative Dual-Mode Dynamic Characteristics

As discussed above, the primary function of the accommodation system is to produce a clear retinal image of objects viewed at various distances. Normal human focusing is accomplished by means of a control process which senses blur of the retinal image and drives the ciliary muscle/lens mechanism to provide a clear image of the object. Accommodative responses to step changes in target distance typically exhibit a latency of 360–380 msec followed by an exponential rise in response with a time constant of about 250 msec.[11,12,152] Although traditionally the accommodation system was considered to be a continuous feedback system, there is an important reason why this is not an appropriate model. Since the response latency is considerably longer than the time constant of the response, its effect in a feedback loop is either to augment the response when the instantaneous error is small or to reduce the response when the instantaneous error is large. For example, as the accommodative movement nears completion with its resultant small instantaneous error, the delayed accommodative drive, reflecting an earlier larger error, becomes effective at this time and overdrives the lens. This mismatch in the length of response latency and time constant has been the source of much difficulty in modeling the accommodation system with continuous feedback control alone.[115]

An indication as to the possible resolution of this difficulty in modeling was provided by the vergence responses to ramp stimuli[148] (see Fig. 21), which indicated a dual-mode process consisting of an open-loop fast component followed by a closed-loop slow component. To determine whether the accommodative system also exhibited dual-mode behavior, a series of accommodative ramp stimuli experiments were conducted.[72] A representative set of accommodative responses to ramp stimuli is shown in Fig. 23. The responses show a progression from ramp (R) response for slow stimuli, step-ramp (SR) and multiple-step (MS) for intermediate velocity stimuli, to step (S) responses for faster stimuli. A histogram of the distribution of response type as a function of stimulus velocity is shown in Fig. 24. Note a shift in the distribution from mostly R responses for stimulus velocity of 0.5 D/sec, to numerous SR responses for 2.5 D/sec, and SR and S responses for step stimuli.

The general shift in the response distribution pattern as ramp stimulus velocity increased revealed an important fundamental dichotomy in the accommodative control processes. A fast control process generated step movements in response to more rapidly moving stimuli ($\geq$ 2.5 D/sec) having a relatively large rate of change of blur, while a slow control process mediated the accommodative responses to the more slowly moving stimuli (< 2.5 D/sec) having a relatively small rate of change of blur. Some investigators have proposed a switching mechanism to account for this behavior.[161]

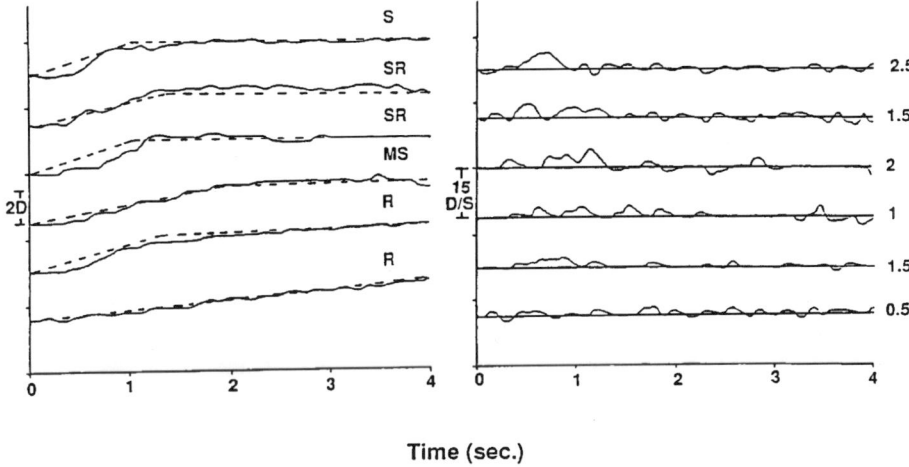

Fig. 23 Representative dynamic accommodative responses for different ramp velocity stimuli (0.5 to 2.5 D/sec) showing progression from mostly ramps (R) to slow velocity stimuli, to step-ramp (SR) and multiple-step (MS) responses for intermediate velocity stimuli, and mostly steps (S) for higher velocity stimuli. On the left are individual accommodative response (solid lines) and stimulus (dashed lines) traces, and on the right are the corresponding velocity traces. Response type is indicated to the right of the response trace. Stimulus ramp velocity is indicated to the right of the velocity trace. Note that in the 3rd trace from the top, the SR response is followed by another step. Sub. GH (reprinted from Hung and Ciuffreda,[72] p. 329, with permission).

The behavior of the fast component could be observed upon closer analysis of the SR and other responses containing steps. These movements were not simply delayed responses to error magnitude, but instead consisted of steps that fell within the main sequence. The amplitude of the initial step movements to ramp stimuli exhibited two important properties. First, they increased with increasing ramp stimulus velocity. Second, the response approximately matched the instantaneous ramp stimulus amplitude, in spite of the 360 msec delay (i.e. latency) in the feedback loop. Thus, it appeared that the neural control process converted the error rate information into step amplitude information. This would involve preprogramming of the necessary step controller signal to generate an open-loop step movement. Then the slow controller took over and reduced the residual error to a minimum using closed-loop feedback control.

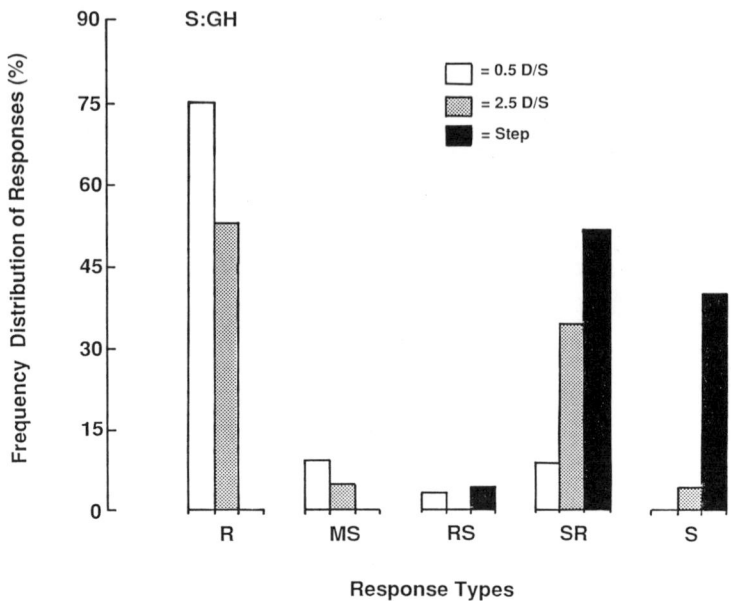

Fig. 24 Histogram showing a shift in distribution from predominantly ramps (R) for 0.5 D/sec ramps, to step-ramps (SR) and ramps (R) for 2.5 D/sec ramps, and mostly SR and steps (S) for step stimuli. Sub. GH (reprinted from Hung and Ciuffreda,[72] p. 330, with permission).

Adaptation Model of Accommodation and Vergence

Most biological systems exhibit the ability to adapt to changing conditions in the environment. Sensory adaptation, such as in the olfactory and auditory systems, serves to shift the operating level over a wide dynamic range. Motor adaptation, such as in muscles, serves to maintain accuracy of movements despite changes in external load. Adaptation is useful, therefore, in balancing the operating level with the demand level of the environment so that the full range of transient responsiveness can be attained.

Adaptation has been observed following prolonged accommodative effort. The accommodation system detects blur of the retinal image of the target and uses visual feedback to change lens curvature until the image becomes clear.[39]

If the accommodative loop is opened, for example in darkness, the lens power decays towards a tonic value, which is subject-dependent and varies between 1 to 4 D.[99,109] Schor[145] examined the effect of different means of opening the accommodative loop on the decay rate. After 5 sec of monocular viewing at 2 D, the rapid decay time constants in the dark ranged from 3 to 5 sec. However, after 60 sec of monocular viewing, the decay rate was much slower and differed according to the means of opening the loop. For example, opening the loop by pinhole viewing, empty field, and darkness for 15 sec resulted in decreases of about 0.2, 0.9 and 2.0 D, respectively.

Similar to accommodation, adaptation was also observed following prolonged vergence effort. If one eye was occluded, thus opening the vergence loop, the angle between the two eyes decreased and decayed toward a tonic value ranging from −1 to +1 MA, with a time constant of ~10 sec.[1,77] However if binocular fusion under prism viewing was maintained for a longer time (e.g. 1–2 min), the subsequent open-loop decay time constant increased manyfold, so that the open-loop vergence level remained close to the binocular fusional angle for a number of minutes. This phenomenon has been called prism adaptation.[55]

An adaptation model was developed by Hung[68] to simulate these accommodative and vergence adaptive responses. The model was based on the static dual-interactive model of the accommodation and vergence system.[77] The approach to the development of the model was to introduce the fewest number of components that would give responses consistent with the experimental findings. In addition, the model was configured so that when it returned to the steady-state condition over time, it became identical to the original static model.

The components of each of the accommodation and vergence sub-systems are similar (Fig. 25). Both contain a controller in the forward-loop. The controller consists of a fast and a slow component. The fast component drives the initial dynamic portion of the response, whereas the slow component maintains closed-loop feedback of the steady-state level. The tonic component represents the value of accommodation or vergence when the feedback loop is opened.[77] This is done by opening the switch in the model. In this way, the controller output goes to zero, so that eventually only the tonic component drives the response. To account for the change in decay characteristics

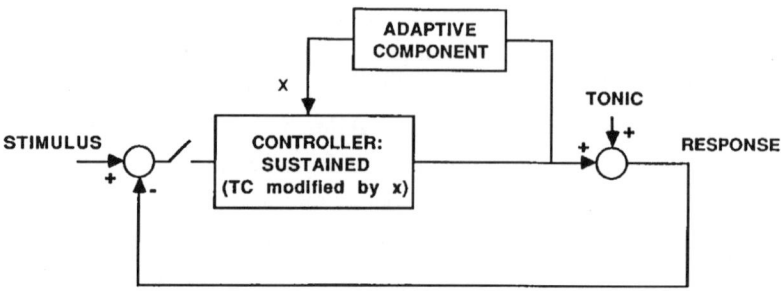

Fig. 25 General adaptation model of accommodation and vergence. For simplicity, the transient element in the controller is not shown. The configuration is basically the same as the general non-adaptation model except the output of the controller also drives an adaptive component. The adaptive component output x modifies the time constant of the sustained element in the controller. An increase in stimulus level increases the controller output, which in turn increases x and thereby the controller time constant. If the switch is opened to simulate the open-loop condition, the controller decays towards zero, but at a much slower rate because of the increased time constant. Eventually the response returns to the tonic level (reprinted from Hung,[68] p. 321, with permission).

following sustained fixation, an adaptive component is incorporated which receives its input from the output of the controller. This is consistent with the concept that adaptation is related to the effort of accommodation or vergence.[67] The unique feature of the model is that the sustained element time constant is modified by the adaptive component output. Although this configuration is unique among near-response oculomotor models, the modification of a component's time constant is seen in other systems. For example, in the saccadic system, Optican and Miles[120] simulated adaptation using modification of time constants. Thus, in the accommodation and vergence sub-systems, as controller output increases with increased effort, the adaptive component increases the time constant of the sustained element. If at that point the feedback loop is opened, the decay of the sustained element will take longer, thereby simulating the longer decay time following prolonged adaptation found experimentally.

The combined accommodation and vergence adaptation model is shown in Fig. 26. Consider first the accommodative loop. The deadspace element (±DSP) represents the neuro-optical DOF. The controller output is multiplied

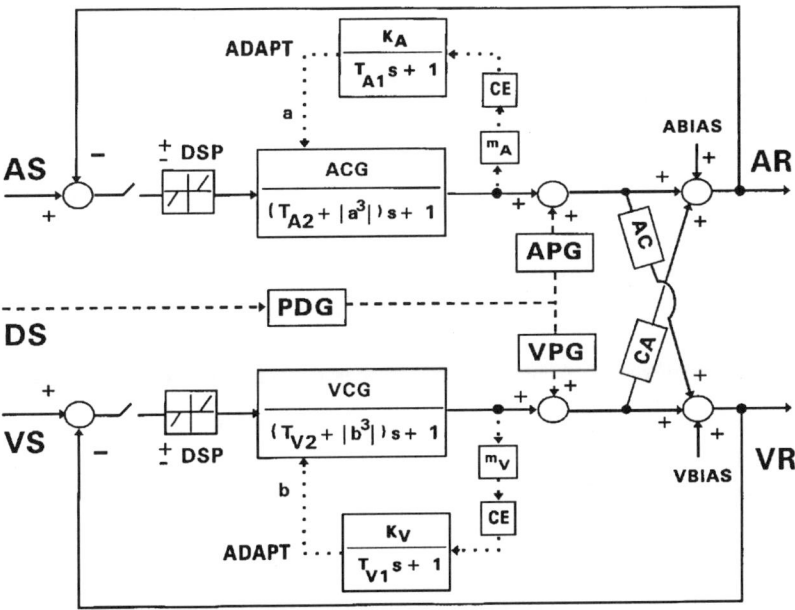

Fig. 26 Adaptation model of combined accommodation and vergence system. The accommodative and vergence controller outputs multiply the factors m_A and m_v, respectively before input to the compression elements (CE), which serve to reduce the effect of large amplitude controller values. In addition, the proximal components, perceived distance gain (PDG), accommodative proximal gain (APG) and vergence proximal gain (VPG) were included for the NITM simulations (reprinted from Hung and Ciuffreda,[73] p. 152, with permission).

by factor m_A and input to a tanh function which serves as a compression element (CE). The factor m_A is used to provide an appropriate range on the abscissa of the tanh function. The CE reduces the controller output for large magnitude inputs so that the adaptation effect is not drastically different at various adapting stimulus levels. The adaptive component is represented by the first-order dynamic element

$$\frac{1}{T_{A1} s + 1} \qquad (29)$$

where T_{A1} is the accommodative adaptation time constant. The accommodative adaptation gain, K_A, controls the magnitude of the adaptive component output level. The adaptive element output, a, modifies the time constant of the accommodative controller via the term, $T_{A2}+|a^3|$, where T_{A2} is the fixed portion of the time constant. The cubic relationship was obtained empirically to provide a negligible increase in time constant for a small amount of adaptation, but very long time constant for a large amount of adaptation.

A similar configuration applies to the vergence system, where the deadspace element (±DSP) represents PFA.[114] The vergence adaptive components consist of multiplier m_v, compression element CE, adaptive gain K_V, adaptive time constant T_{V1}, adaptive element output b, and controller time constant $T_{V2}+|b^3|$.

A number of adaptation experimental results[55,150] were accurately simulated using parameter values listed in Table 2. One example is the effect of different amounts of adaptation on vergence response using the Henson and North paradigm.[55] Initially, both feedback loops were closed with AS = 0.25 D and VS = 4 MA. After 15 sec, the vergence loop is opened for another 15 sec. The cycle was then repeated. The vergence response for normal, abnormal and non-adaptive conditions were simulated using progressively decreasing saturation levels for the controller time constant[68] (Fig. 27). Note

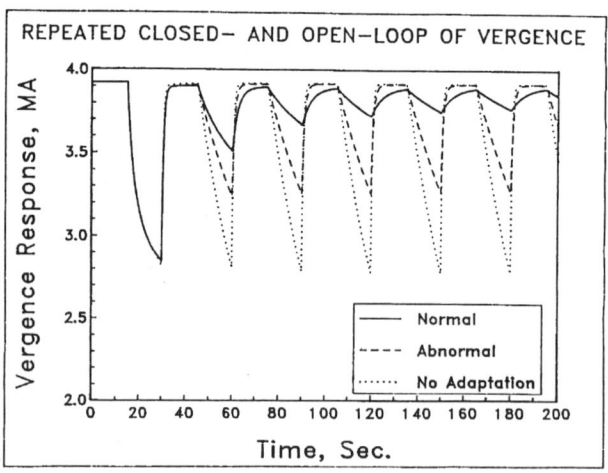

Fig. 27 Model simulation of different amounts of adaptation using the Henson and North (1980) paradigm. Vergence responses are shown for normal (——), abnormal (- - -) and non-adaptive (...) conditions (reprinted from Hung,[68] p. 324, with permission).

that the bottom of the decay curve increases toward the vergence stimulus level over time for the normal, to a lesser extent for the abnormal, and not at all for in the case of no adaptation. The amount of change of the phoria value starting from the first cycle is ~ 1.07, 0.50 and 0 MA, for normal, abnormal and non-adaptive, respectively.

The adaptation simulation results can also explain how accommodative hysteresis occurs.[8,43] Following sustained *near* viewing, the adaptive element output remains high as it begins its decay towards the tonic level, or ABIAS (see Fig. 8B). Hence, immediate post-task accommodative response measured in the dark is a higher than the tonic value. On the other hand, following sustained *far* viewing, the adaptive element output remains low as it begins to rise towards the ABIAS value, and the immediate post-task response is a lower than the tonic value. This dependence of the post-task accommodative value on the path, or initial condition (either near or far sustained viewing), gives rise to the phenomenon of accommodative hysteresis.

Nearwork-Induced Transient Myopia (NITM) Model

Myopia is a clinical refractive condition that affects 25% of the adult population in the United States[157] and at least 75% of the population in Asian countries such as Taiwan.[100] Myopia can be corrected by optical means, but the cost to the consumers in the US for eye examinations and optical corrections is US$4.6 billion per annum.[85] Furthermore, the wearing of spectacles for myopia may restrict one's vocational options.[102] Surgical techniques have been developed over the past 20 years to correct myopia,[85] but they are expensive and may have side effects. Moreover, they do not prevent the subsequent development of adult-onset myopia or other age-related refractive changes. Thus, myopia is a costly worldwide public health problem. For these reasons, the slowing of myopic progression, as well as the prevention of its initial occurrence, has been of considerable interest to clinicians and scientists alike. Yet, a deeper understanding of the underlying myopigenic mechanisms has only recently begun to evolve.[30,50,116,117,138]

The development of myopia has both genetic and environmental components.[48] Although genetic factors appear to play a larger role in early-onset myopia, modern-day workstyles clearly demonstrate that environmental factors may play a significant role in the development of late-onset myopia.[116,117] A particularly important environmental factor is that of prolonged nearwork, which has been especially implicated in the development of late-onset myopia (i.e. onset after 15 years of age).[116,117]

Four refractive groups have been identified in terms of severity (and direction) of refractive error: hyperopia (HYP), emmetropia (EMM), early-onset myopia (EOM), and late-onset myopia (LOM).[50] Quantitative measures would be helpful to differentiate between, and perhaps even predict, those who will develop myopia versus those who will not. Such a measure can be found by stimulating the accommodation system during near viewing, which produces lenticular-based pseudo-myopia, and then measuring the closed-loop temporal course of decay of the lens response back to the original far point of accommodation. This is referred to as the nearwork-induced transient myopia (NITM) paradigm,[36,96,116,117,135] with the difference between post- and pre-task values representing the NITM.

The adaptation model[68] discussed above quantified the effect of prolonged nearwork on the accommodative response. It accurately simulated experimental results on the dynamic accommodative behavior following prolonged nearwork, as well as during alternate nearwork and distance viewing.[18,55,150] This model served as the basis for the simulation of NITM dynamics.[73] An important parameter in the model, the adaptive gain K_A, was previously used to modify the time constant of the accommodative controller, and thus controlled the rate of decay in the dark following an adaptation period. Simulations were therefore performed to determine whether variation in the K_A could also account for the differences in the dynamic decay timecourse in the light following nearwork in the different refractive groups. In addition, the computer-simulated effect of higher K_A values on retinal defocus was examined over a 160 hr period, representing one work-month with 40 hours of nearwork per week, to assess its influence on the long-term development of myopia.

To provide a more complete nearwork model for the NITM simulations, proximal elements were added to the adaptation model. The input to the proximal component was represented by a distance stimulus (DS), which drove the perceived distance gain (PDG). The output of PDG was input to both the accommodative proximal gain (APG) and vergence proximal gain (VPG) elements, which were summed with the respective controller outputs (Fig. 26). It has been shown that while the proximal component constituted a considerable percentage (up to about 80%) of the accommodative response under open-loop conditions, it provided a negligible contribution (<4%) under normal closed-loop conditions.[74] Thus, under the closed-loop paradigms used in the present modeling study on NITM and long-term accommodative error, the proximal component is expected to play a very minor role. The proximal component is included in the model, however, for completeness of our new comprehensive nearwork model.

In contrast to the proximal component, the accommodative adaptation gain played a crucial role in inducing NITM due to its effect on the accommodative controller time constant. That is, after sustained nearwork, the increased accommodative adaptation element output would result in an increase in controller time constant. This in turn would result in a slower than normal return or decay of the NITM toward the pre-task baseline level. Larger adaptive element gains produced slower decay rates, so that different gain values could be used to simulate NITM in the various refractive groups. The parameter values used were based on the adaptation and proximal models[68,74] (Tables 2 and 4).

The nearwork model was used to simulate the NITM response (2 min at far; 0.17 D and 0.17 MA) following 10 min of congruent binocular near viewing at 20 cm (5 D and 5 MA). In the simulation runs, the K_A value was varied from 1.0 to 6.0 in increments of 0.5. The NITM simulation curves were compared to and matched visually with the experimental results (Fig. 28A) for the four refractive groups. It was found that the K_A values of 2.0, 2.5, 4.0 and 5.5 simulated reasonably accurately the experimental NITM time courses for HYP, EMM, EOM and LOM, respectively (Fig. 28B). For the simulation of the response of the HYP group, an additional constraint was imposed wherein the accommodative response to distant stimuli was biased on the under-accommodated side of the deadspace operator (or DOF). This was done for consistency with the experimental results.[25] In contrast, for the other three refractive groups, no such constraint was imposed so that the accommodative responses exhibited the normal slight (~0.25 D) over-accommodation for the far target.[136]

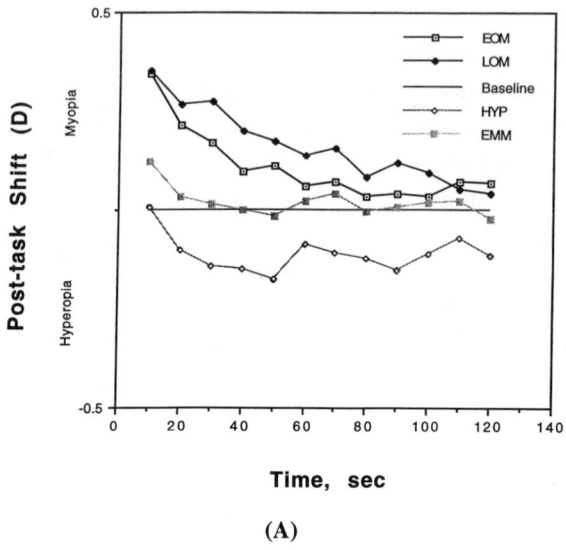

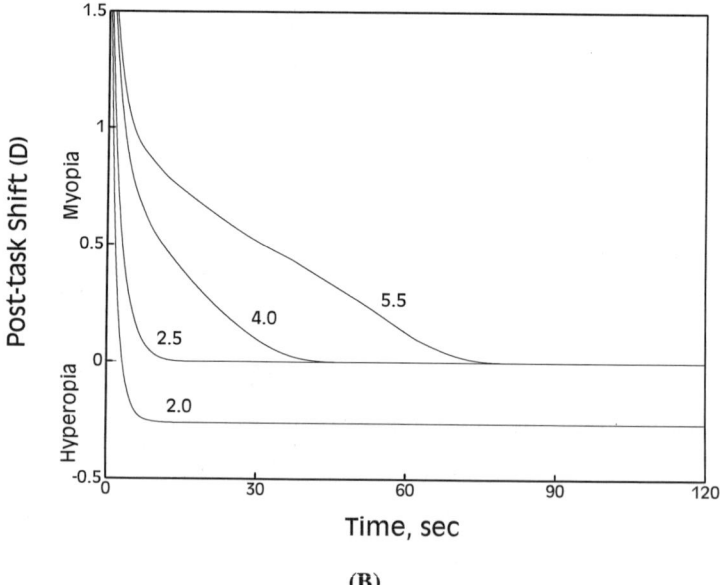

Fig. 28 (A) Experimental results (pre-task baseline distance refraction variability was < ± 0.1 D), and (B) matching model computer simulation of NITM as a function of refractive group (reprinted from Hung and Ciuffreda,[73] pp. 155–156, with permission).

In addition, the model was used to investigate long-term effects of nearwork. The paradigm consisted of alternating periods of prolonged nearwork and brief distance viewing, which is representative of our everyday activities. Under this condition, both under-accommodation at near (lag of accommodation) and over-accommodation at far (lead of accommodation) typically occur.[16,25] A useful measure of the long-term effect of the resultant retinal defocus on an individual, regardless of how it is generated, is that of the root-mean-square (rms) of the accommodative error. The rms error is essentially equal to the standard deviation about the mean value, so that a larger value is associated with greater retinal defocus. This measure was used in the prolonged nearwork simulation. This was simulated by alternating 1 hr of congruent nearwork (3 D, 3 MA) and 5 min of congruent far viewing (0.25 D, 0.25 MA) over a 160 hr period, which represents one work-month with 40 hours of nearwork per week. The final steady-state rms value of the overall (i.e. combined for distance and near conditions) accommodative error was measured and plotted as a function of K_A (Fig. 29). The results show a

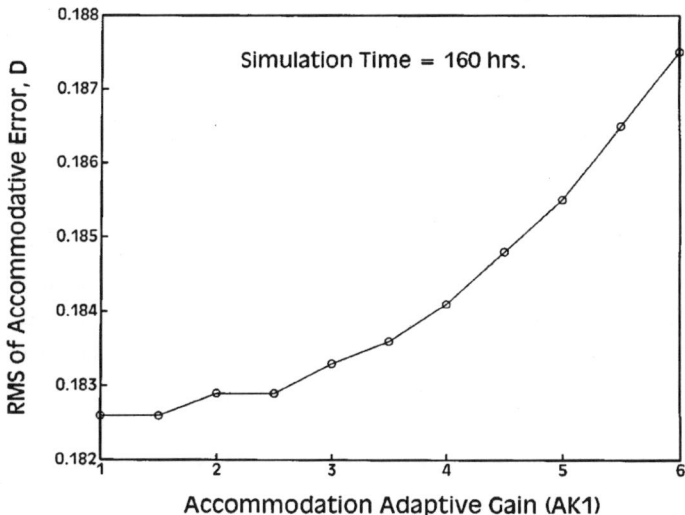

Fig. 29 Model computer simulation showing effect of accommodation adaptive gain, K_A, on the rms value of accommodative error (reprinted from Hung and Ciuffreda,[73] p. 156, with permission).

small but progressive increase in the rms of the accommodative error with increased K_A.

Genetic factors may influence the link between K_A, defocus blur, and myopia development. These factors include both neuro-adaptive and biomechanical aspects. For example, individuals may have a genetic predisposition to greater adaptive response during near viewing, resulting in transient over-accommodation at distance. This is seen as a slower accommodative decay of NITM that can eventually lead to an excess of cumulative, time-integrated retinal defocus, perhaps reflecting slightly impaired sympathetic activation of the ciliary muscle.[160] Individuals may also have a greater susceptibility and biomechanical compliance of the ocular tunics to retinal defocus effects. Defocus effects on the retina may in turn induce biochemical changes that result in an increase in axial length and hence myopia.[48] The differential expression of these two genetic factors may be responsible in part for the variation of the degree of myopia found.

Environmental factors can act conjointly with an individual's genetic predisposition, serving as a trigger mechanism for the manifestation of the genetic component, or perhaps they are simply additive. Thus, it may be possible for an individual to eschew nearwork, and the associated temporal increase in retinal defocus, to avoid or reduce the development of myopia, despite a familial genetic predisposition for myopia. Conversely, an individual may perform an over-abundance of nearwork, thus greatly increasing exposure to chronic retinal defocus, resulting in the development of myopia despite the lack of such a familial genetic predisposition.

The emphasis of this model-based study was the role of the adaptive component on the NITM timecourse. Indeed, the NITM simulation results demonstrated that a single parameter, the accommodative adaptation gain K_A, clearly discriminated the four refractive groups. The lower adaptive gain values (K_A = 2.0 and 2.5) corresponded to the HYP and EMM groups, respectively, whereas the higher adaptive gain values (K_A = 4.0 and 5.5) corresponded to the EOM and LOM groups, respectively. The adaptive element represented a neural-oculomotor feedback process that controlled the effect of sustained stimulation of the accommodative system during nearwork. The slowed decay of the distance accommodative response following prolonged nearwork was due to an increase in output from the adaptive element, thus resulting in an increase in the accommodative controller time constant. This decay was slower for larger K_A values. The long-term increase in exposure to accommodative error, and the resultant retinal defocus, was simulated using repeated periods of prolonged nearwork and brief distance viewing, representing rest intervals. This provided a differentiation among the refractive groups. The effect of NITM on the development of myopia is discussed in the next section.

Refractive Error Development Model

Background

If the cornea and lens grow in concert with the scleral tunic during the ocular growth and development phase, there would be a perfect match between the refractive (i.e. optical) and mechanical size components of the eye, i.e. emmetropia. Thus, genetic control of growth plays an important role in the development of normal vision. However, environment factors, such as prolonged nearwork, have also been shown to have a significant effect on the deviation from normal growth and the consequent development of refractive error, in particular myopia.[116,117] Moreover, differences in individual susceptibility to nearwork effects during growth and development add to the complex interplay between genetic and environmental factors in myopia development; for example, young adult myopes are more susceptible to nearwork accommodative aftereffects than either emmetropes or hyperopes.[25] Therefore, understanding the underlying mechanisms of myopia development and eventually its prevention has been of considerable interest in recent years.[4,166]

Recall from the previous section on "Static Analysis Techniques" that due to the depth of focus as well as accommodative control gain, the accommodative response is typically less than the stimulus for near viewing (i.e. "lazy" lag of accommodation), so that the retinal image is focused slightly behind the retina[15,16] (Fig. 5). Recent evidence in animals has shown that during development, the eye grows in the direction of this induced retinal-image defocus.[144,154] This suggests that the eye may elongate in response to the resultant blurred image as part of the emmetropization process (i.e. the tendency for the eye to grow and become more emmetropic when presented either with hyperopic or myopic blur). Indeed, there is evidence that children who become myopic have relatively inaccurate accommodation at near to blur-only stimuli.[51] Blur-induced biochemical factors appear to play a role in causing local scleral growth,[14] since this occurs despite severing of the optic nerve.[165] Thus, it appears there are two primary factors that control ocular development: normal genetically programmed growth of the refractive elements and the sclera, and contribution of local retinal-image defocus along with a susceptibility factor to either enhance or retard the growth rate.

A previous model of emmetropization in the chicken that separated the optical and retinal-defocus processes was developed by Schaeffel and Howland.[143] They proposed that the experimental data obtained could be described by processes containing two feedback loops. One loop was dependent on lens response feedback, and the other loop was dependent on retinal-image feedback. They showed model simulation fit to some of their experiment data. However, this model was simplified in that it ignored interactions between the two loops. Medina and Fariza[105] proposed a model of emmetropization in which the overall transfer function is given in the form $F(s) = 1/(1+ ks)$, where the input is the command refractive error and the output is the measured refraction. However, to fit his refractive error versus age experimental curve, the time constant k would have to be very large (about 10 years). The problem then is that the model cannot respond rapidly (in seconds) to step changes in accommodative stimulus seen in normal individuals. Flitcroft[44] proposed a model of emmetropization and myopia in humans based on the dual-interactive model of Hung and Semmlow.[77] He used an integrative measure of accommodative error that was dependent on the fraction of time spent at near and at far. This was used to update the refractive error. Model simulations showed a reduction of refractive error following a number of iterations that simulated emmetropization, and the development of myopia following prolonged nearwork. The simulations showed however, an increase in refractive error after wearing corrective spectacles as compared to no spectacles. Moreover, no genetic control factors or thresholds that accounted for different susceptibility in various refractive groups were used. Thus, these models have not been able to explain the fundamental underlying mechanism of refractive error development.

To explore the underlying mechanism and to account for the experimental findings, a recent theory of refractive error development was constructed.[69] Two fundamental insights underlie the theory, called the Incremental Retinal-Defocus Theory. First, local retinal-defocus *magnitude* is critical in the development of environmentally-induced refractive error. Second, manipulations of the optical environment are effective in producing refractive error mainly during the ocular growth and maturational period (in contrast, most mature adult animals are relatively insusceptible to optical manipulations[48]). Each insight alone is *insufficient* to provide a workable bidirectionally-sensitive theory, but when these two are combined, they provide a coherent framework for a unifying theory of refractive error development. The theory is based on the concept that the *change* in magnitude of retinal defocus during an *increment* of genetically programmed axial length growth provides the critical information for directional modulation of growth rate (describe in detail below). The term "genetically programmed" is used to describe the normally occurring ocular growth that has been preprogrammed genetically, which should be distinguished from environmentally induced growth that is due to change in retinal defocus. Both involve neuromodulator release, with the environmentally induced component in effect modulating the normal genetically programmed release rate. This theory was tested under five critical experimental conditions,[69] two of which are presented below, to assess the generality of the proposed underlying mechanism. In addition, a MATLAB/SIMULINK model was constructed to quantitatively demonstrate the direct effect of change in retinal-defocus, via signal cascade through the retinal layers, on scleral growth rate.

Incremental retinal-defocus theory

The development of a theory for the control of ocular growth rate requires the conceptual linking of a number of areas including: cornea and lens growth, optics of the eye, retinal-neural signal processing, scleral growth and ocular neurochemistry.[116] The fundamental principles can be understood in terms of answers to three critical questions: (1) What is the contribution of the cornea and lens to the emmetropization process?; (2) How do retinal neurochemicals process the retinal-defocus information?; and (3) How is this information used to regulate the rate of ocular growth?

(1) Corneal growth does not contribute to the emmetropization process after two years of age

During the first two years of life, the cornea and axial length overcome the initial hyperopia and grow rapidly and in concert as part of the emmetropization process.[142] Thereafter, corneal power remains relatively stable,[48,156] until the adult years, when it may increase slightly.[50] Since corneal flattening rather than steepening would be needed to compensate optically for the axial length growth, it is clear that the cornea plays little or no role in the emmetropization process after the first two years of life.[48,156] Moreover, since there is no evidence that visual feedback plays a role in the growth of the lens,[156] emmetropization during this later period involving any large artificially induced retinal-image defocus must be provided only by the rate of change of axial length growth.

(2) Neuromodulators control sensitivity to changes in retinal-image contrast

In contrast to neurotransmitters (such as glutamate, acetylcholine and GABA) which respond rapidly to retinal stimulation, neuromodulators (such as dopamine, seratonin and neuropeptides) act over a longer period, and may also cause changes in the neuronal synapses.[111] An example of synaptic plasticity in the retina can be seen in the interplexiform cells in the retina.[111] These neurons (which contain dopamine) receive their inputs from the amacrine cells in the inner plexiform layer, and then send their outputs back to the horizontal cells in the outer plexiform layer. Dopamine serves as a neuromodulator by altering the properties of the horizontal cell membrane and decreasing the flow of current across the membrane. Moreover, because of the center-surround structure of the retina, the interplexiform neurons respond in a graded manner to local retinal-image contrast.[111]

The theory proposes that feedback regulation provided by the interplexiform neurons from the inner to outer plexiform layers maintains a relatively constant sensitivity to retinal-image contrast.[69] Such feedback regulation is useful because it would preclude the need for a memory mechanism to register and store previous levels of retinal-defocus for the purposes of update and comparison. The release of neuromodulators by the interplexiform neurons result in synaptic changes in the horizontal cells.[111] This in turn alters retinal sensitivity to center-surround input, which helps to shift the steady-state operating level to permit responsivity to transient changes in local retinal-image contrast. Thus, the net rate of release of neuromodulators is not dependent on the absolute level of retinal defocus, but rather on the *change* in retinal-defocus magnitude. The release of neuromodulators also causes structural changes in the sclera via modulation of proteoglycan synthesis;[113] wherein an increase in proteoglycan synthesis rate results in a greater structural integrity of the sclera, and in turn, a decrease in axial growth rate relative to normal; conversely, a decrease in proteoglycan synthesis rate results in less structural integrity of the sclera, and in turn, an increase in axial growth rate relative to normal.[104,113,166]

(3) The overall mechanism for regulating the rate of axial length growth

Genetically programmed mechanisms determine a baseline rate of neuromodulator release that is associated with the normal axial growth rate. Retinal defocus-induced changes in the rate of neuromodulator release are superimposed onto this baseline level to result in changes relative to the normal axial growth rate. The net effect of the local-retinal mechanism, as discussed above, is that the change in retinal-defocus magnitude, and in turn the change in the rate of neuromodulator release, are in *opposite directions* with respect to the change in the rate of defocus-induced axial growth relative to normal. Thus, during an increment of genetically programmed ocular growth, a change in retinal-defocus magnitude due to the incremental change in ocular geometry provides the directional information needed to modulate the rates of release of neuromodulators and proteoglycan synthesis, which in turn produce structural changes in the sclera for regulation of ocular growth.[113] For example, during an increment of genetically programmed ocular growth (over days), if the retinal-defocus magnitude decreases, the axial growth rate increases. This results in relative myopic growth. On the other hand, if the retinal-defocus magnitude increases, the axial growth rate decreases. This results in relative hyperopic growth. These resultant axial growth rate changes are consistent with the emmetropization process.

Applications of the theory

Lenses

During ocular development, the eye exhibits continuous genetically programmed growth. The imposition of a lens causes changes in retinal-defocus which act to modulate the genetically predetermined normal growth rate, and thereby alters the overall axial growth rate. This modulation can be illustrated by the following example. Consider the effect of introducing spherical lenses in front of the eye. The change in size of the blur circle during a *small increment* of normal genetically programmed ocular growth for large imposed zero, minus and plus lenses is shown schematically in Figs. 30A, B and C, respectively. A neuromodulator (such as dopamine) maintains a certain level of neuronal activity related to retinal-image contrast by means of the local retinal feedback mechanism described earlier. The net effect is that the rate of neuromodulator release is dependent not on the absolute level of retinal-defocus magnitude, but rather on the change in retinal-defocus magnitude during the increment of genetically programmed ocular growth, as was also mentioned earlier. For example, for a zero power lens, there is no change in the size of the blur circle. Thus, no additional neuromodulator is released, and the normal genetically based incremental axial growth pattern of the young eye is maintained. With the introduction of a minus lens however, the size of the blur circle is decreased, thus the rates of neuromodulator release and in turn proteoglycan synthesis are decreased, thereby resulting in an increase in axial growth rate.[113] On the other hand, with the introduction of a plus lens, the size of the blur circle is increased; thus, the rates of neuromodulator release and in turn proteoglycan synthesis are increased, thereby resulting in a decrease in axial growth rate.[113] Hence, either a decrease or increase in mean retinal-defocus magnitude during an increment of genetically programmed axial growth is proposed to cause a change in the rate of neuromodulator release, which in turn leads to structural changes in the sclera,[113] that are manifest as appropriate changes in the rate of axial growth and reflect the active emmetropization process.

Change in Retinal Defocus During Increment of Normal Genetically-Programmed
Axial Length Growth for Different Imposed Lenses.

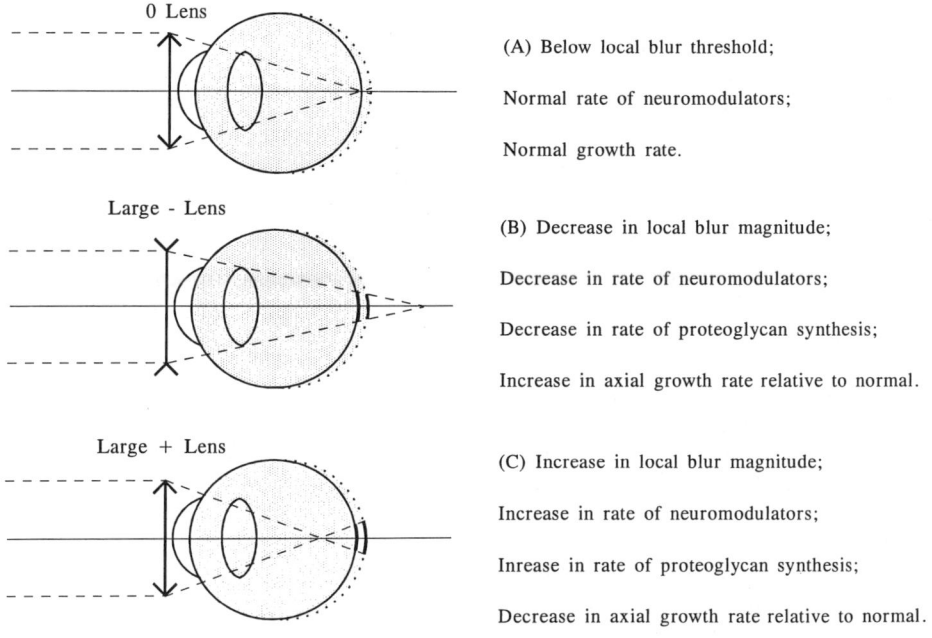

(A) Below local blur threshold;

Normal rate of neuromodulators;

Normal growth rate.

(B) Decrease in local blur magnitude;

Decrease in rate of neuromodulators;

Decrease in rate of proteoglycan synthesis;

Increase in axial growth rate relative to normal.

(C) Increase in local blur magnitude;

Increase in rate of neuromodulators;

Inrease in rate of proteoglycan synthesis;

Decrease in axial growth rate relative to normal.

Fig. 30 Schematic representation of change in blur circle during a small increment of normal genetically programmed ocular growth [From initial time (solid) to a short time later (dashed)] under the conditions of: **(A)** zero lens; **(B)** minus lens; and (C) plus lens (reprinted from Hung and Ciuffreda,[69] with permission).

Prolonged nearwork

The Incremental Retinal-Defocus Theory can be applied to the condition of prolonged nearwork, which has been implicated in the development of school myopia, wherein relatively small amounts of retinal defocus are present over extended periods of time (i.e. weeks or months).[116,138] This can be understood by examining the normal accommodative stimulus/response (AS/R) function[15,16,19,66,118] (Fig. 31). This function is an *s*-shaped curve showing slight over-accommodation at distance and progressive under-accommodation at near with increased dioptric demand. Thus, during nearwork, which is represented

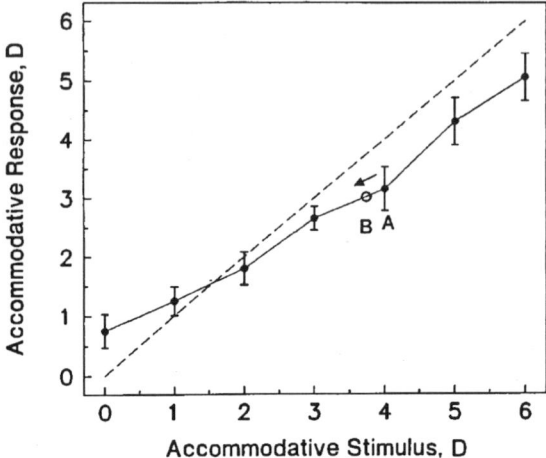

Fig. 31 Plot of averaged accommodative stimulus-response data for ten visually normal subjects. Filled circles and error bars represent group mean ± SEM (adapted from Ong et al.,[118] with permission from the authors).

by a relatively large accommodative stimulus, the accommodative response lags the stimulus which results in chronic hyperopic defocus (e.g. pt. A in the AS/R curve, Fig. 31; Fig. 32A). Following nearwork, viewing at far (Fig. 32B) results in a small transient myopia (NITM). Also, due to the slowed decay of crystalline lens power, a subsequent return to nearwork will exhibit a far-to-near carry-over of NITM, which can be represented schematically as an equivalent plus lens (Fig. 32C). Due to the presence of the added equivalent plus lens, the effective AS is now reduced slightly, so that AR moves down from A to B in the AS/R curve (see Fig. 31). Thus at point B, less accommodative response would now be necessary for clear retinal imagery. This also means the accommodative error, or retinal defocus, is slightly reduced. And according to the same arguments above regarding large imposed minus lens (Fig. 30), there would be a slight decrease in the rates of neuromodulator release and proteoglycan synthesis. This in turn would result in an increase in axial growth rate relative to normal, i.e. myopic growth, which is consistent with the emmetropization process. Moreover, if the retinal-defocus is left uncorrected over a prolonged period of time, the process would repeat itself and effectively continue to move slightly down and to the left on

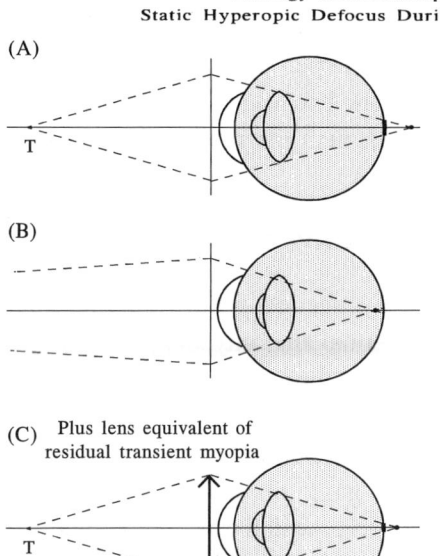

Fig. 32 Schematic representation of the effect of repeated NITM on change in retinal defocus (adapted from Hung and Ciuffreda,[69] with permission).

the AS/R curve, which would eventually lead to continued increase in axial growth rate relative to normal, and in turn a progressive development of myopia.[116] This process is similar to that for the imposition of a minus lens (Fig. 30), except in this case it is the shape of the AS/R curve[66,118] (i.e. the decrease in the difference between the curve and the 1:1 line as AS decrease) (Fig. 31), rather than genetically programmed axial growth *per se*, that drives the increase in axial growth rate and subsequent myopia development. It is somewhat ironic that, myopia development, rather than being a failure of the emmetropization process, is actually a result of an emmetropization process that operates under the constraints of the relationship between the AS/R function and the 1:1 line.

The above analysis also explains why the increase in axial growth rate is primarily associated with prolonged near- rather than far-work. The reduction in accommodative error during an increment of genetically programmed ocular growth is more pronounced for near (>3 D) than for far (<1 D) viewing (see Fig. 31). Furthermore, some individuals (hyperopes and emmetropes) may not develop myopia subsequent to nearwork because they may have higher thresholds for inducing axial length growth than the myopes.[69]

Basic retinal anatomy and physiology

To understand the retinal signal pathway needed within an Incremental Retinal-Defocus model, a brief review of retinal anatomy and physiology may be helpful. Signals are transmitted in the retina through three types of neurons: photoreceptors, bipolar cells and ganglion cells. The photoreceptors (rods and cones) are stimulated by light and relay the signal through bipolar cells, which in turn relay the information to the ganglion cells. The axons of the ganglion cells in the retina form the optic nerve, which transmits retinal-image information to the higher cortical centers. Bipolar cells also receive light stimulus information from neighboring or surround photoreceptors, via lateral connections from horizontal cells in the outer plexiform layer. This center-surround organizational structure provides local retinal-image contrast information to the "sustained" ganglion cells, which respond to sustained contrast information. On the other hand, "transient" ganglion cells respond to *change* in the surround via amacrine cells in the inner plexiform layer. Thus, these neurons relay information regarding the change in retinal-defocus magnitude. In addition, interplexiform neurons from the inner to the outer plexiform layer modulate the long-term sensitivity of horizontal cells to surround input. Thus, this feedback mechanism serves to adjust the steady-state sensitivity level to provide relatively constant sensitivity to changes in local contrast.[111]

Model of refractive error development

A conceptual block diagram of the Incremental Retinal-Defocus model is shown in Fig. 33A. It is based on the principle that the magnitude of retinal-defocus can be represented by the difference in center and surround excitation. A *change* in this signal, and thus retinal-defocus magnitude,

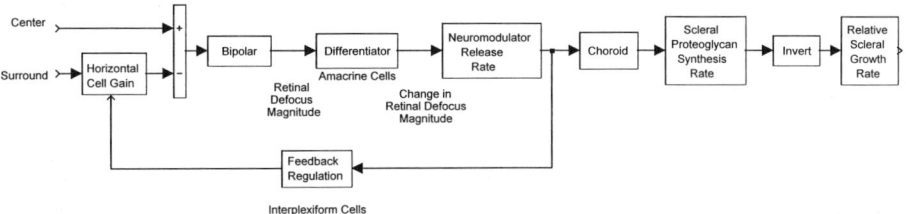

Fig. 33A Conceptual block diagram model of the retinal-defocus pathway for regulating sclera growth. The difference between the center and surround excitation provides the retinal-defocus signal. The derivative of the signal drives the release of neuromodulators, which provides the feedback via interplexiform neurons to regulate horizontal cell gain. In addition, release of neuromodulators cause changes in the rate of proteoglycan synthesis and in turn, relative scleral growth rate (reprinted from Hung and Ciuffreda,[69] with permission).

provides the requisite sign for modulating ocular growth. The sensitivity to local retinal-image contrast is maintained at a relatively constant level by means of feedback regulation of horizontal cell gain provided by the interplexiform neurons. This precludes the need for a "memory mechanism" for storing information regarding the previous level of retinal-defocus magnitude, so that its change can be discerned. The release of neuromodulator in turn results in changes in the rate of scleral proteoglycan synthesis, which causes a change in scleral growth rate. This relative growth rate is added to the ongoing normal genetically programmed ocular growth rate to provide the overall axial length growth.

A detailed Incremental Retinal-Defocus model, constructed in the MATLAB/SIMULINK environment, is shown in Fig. 33B. It consists of two pathways — sustained and transient. The sustained pathway consists of the photoreceptor, bipolar and sustained ganglion cells. It is modulated by surround signals via horizontal cells in the outer plexiform layer to provide local steady-state or sustained contrast information. The transient pathway consists of photoreceptor, bipolar and transient ganglion cells. It is modulated by surround signals via amacrine cells in the inner plexiform layer to provide information regarding change or transients in local contrast. Feedback regulation is provided by the interplexiform neurons that receive signals for neuromodulator release in the inner plexiform layer and modulate the gain of horizontal cells in the outer plexiform layer to maintain a relatively

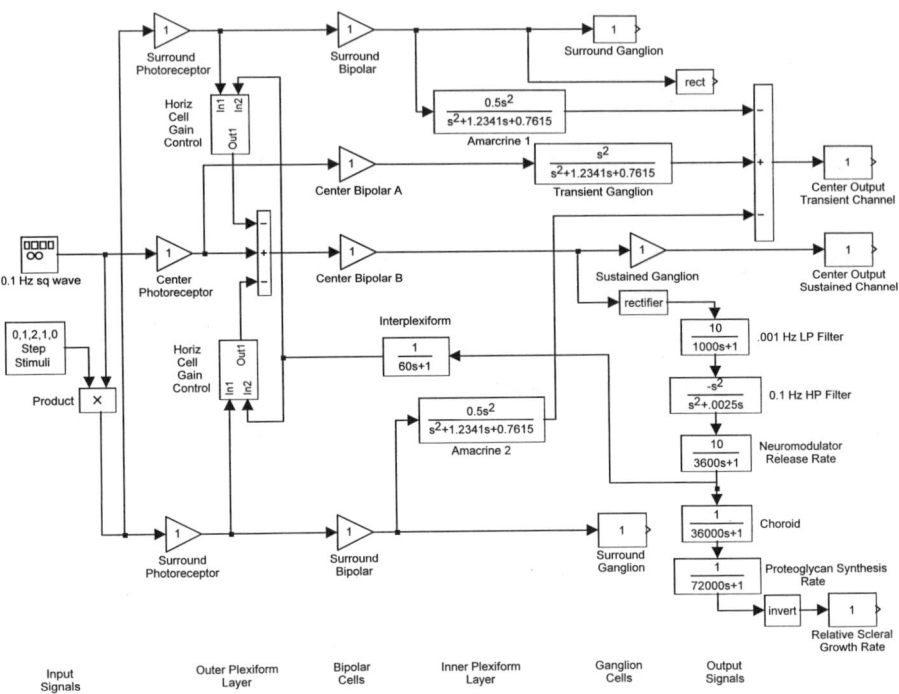

Fig. 33B Detailed block diagram model depicting the regulation of scleral growth rate. The retinal layers (outer to inner) are arranged from left to right: photoreceptor, outer plexiform, bipolar, inner plexiform and ganglion. The sustained pathway consists of center photoreceptor, center bipolar B and sustained ganglion cell. Horizontal cell, whose gain is regulated by feedback via interplexiform cells, relay surround information to modify sustained ganglion output. The transient pathway consists of center photoreceptor, center bipolar A and transient ganglion cell. Amacrine cell relay change in surround information to modify transient ganglion output. Center bipolar B signal consists of retinal-defocus information and passes through a rectifier, lowpass filters and elements representing neuromodulator release, the choroid and proteoglycan synthesis. This is inverted to provide relative scleral growth rate relative to normal (reprinted from Hung and Ciuffreda,[69] with permission).

constant sensitivity to change in local contrast. The center bipolar cell receives a signal derived from the difference between center and summed surround inputs, which thus represents the summated amount of retinal-image defocus across the overlapping, spatially contiguous center and surround receptive field area. This signal is differentiated by a neural circuitry in the inner plexiform layer, which most likely contains amacrine cells. This change is rectified, so that the "envelope" of the signal, which represents the overall change in retinal-defocus magnitude, drives the rate of neuromodulator release. The neuromodulator, or a cascade of neurochemicals related to the release of the neuromodulator,[166] passes through the choroid to reach the sclera. The transit of the neuromodulator through the choroid may result, at least in the monkey, in a volume change that is observed as a change in choroidal thickness.[69,116] This may explain why, as expected, the choroidal thickness changes in the monkey are correlated with changes in retinal-defocus magnitude, but the optical change associated the thickness change is too small to account for any significant contribution towards full emmetropization.[69,116] On the other hand, the neuromodulator that reach the sclera modifies proteoglycan synthesis to result in changes in ocular growth that does provide nearly full emmetropization, as described in the schematic model above.

A simulation was performed using the relative amounts of center and surround intensity input to represent the amount of retinal-defocus. The stimulus amplitude was defined in terms of its relative brightness over a unit of retinal area, in which a unit area represents the extent of the limit of visual acuity (about 1 min of arc linear dimension).[111] For simplicity, each center and surround unit of retinal area could be assigned three brightness levels: -1, 0 or 1. The contribution from an additional unit of spatially extended surround area could be included by adding its brightness contribution from the immediately adjacent surround area. In this way, the surround amplitude reflected the relative amount of retinal-defocus rather than retinal-image contrast *per se*. To simulate the temporal variation in brightness over a particular retinal locus in the course of a normal viewing period, all retinal areas received a 0.1 Hz square-wave signal. Thus for example, the receptive field center would always consist of a ±1 amplitude peak-to-peak (ptp) signal (i.e. a 0.1 Hz square-wave with amplitudes ranging from -1 to $+1$). On the other hand, the surround signal could vary depending on the amount of modulation (0, 1 or 2), which represents the amount of retinal-defocus. For example, a modulation of 0 (i.e. zero amplitude for the surround) represents a relatively small amount of retinal-defocus, i.e., a relatively sharply focused image. A modulation of 1 from one unit area of surround (i.e. a 0.1 Hz square-wave with amplitude ranging from -1 to $+1$) represents a moderate amount of retinal-defocus. Finally, a modulation of 2, due to the summing of two adjacent unit surround areas (i.e. a 0.1 Hz square-wave with amplitudes ranging from −2 to +2) represents a relatively large amount of retinal-defocus.

The quantitative model was tested using a selected series of surround modulation steps (0, 1, 2, 1, 0) at 100 hr intervals over a 500-hr period. This was chosen to demonstrate transitions between various levels of retinal-defocus magnitudes. More importantly, the model simulation was used to demonstrate the long-term effects of changes in retinal-defocus magnitude on scleral growth rate. Various model parameters were monitored: the horizonal cell gain (representing feedback modulation from inner to outer plexiform layer by interplexiform neurons), the rates of neuromodulator release and proteoglycan synthesis, and the relative change in axial length.

Model simulation responses to center and surround stimuli are shown in Figs. 34A–D. The center stimulus, representing sharp focus, consists of a ±1 amplitude ptp, 0.1 Hz, square-wave signal. The surround stimuli, representing varying degrees of retinal-image defocus, consists of the same square-wave but modulated by different step levels over the time span of the simulation. Figure 34A shows the various steps of modulation of the surround amplitude (solid) and the feedback-regulated change in the gain of the horizontal cells. The pulse-like responses for the rates of neuromodulator release (solid) and proteoglycan synthesis (dashed) are shown in Fig. 34B. The change in proteoglycan synthesis rate in turn causes changes in the scleral growth rate relative to normal (Fig. 34C). Finally, the cumulative change in axial length relative to normal is shown in Fig. 34D. These results indicate that the model is able to simulate the bidirectional aspects of choroidal and scleral axial length changes found experimentally.

The schematic analysis above has provided a relatively simple and logically consistent explanation for the eye's ability to grow in the appropriate direction following a lens-induced change in retinal-image contrast. It has also been able to explain the effects of imposed lenses and prolonged nearwork, as well as diffusers, black occluder and removal of the crystalline lens[69] (not presented here) on ocular growth. The critical point is that the detection mechanism does not depend on the sign of the blur, but rather on the change in blur magnitude, such as during genetically programmed ocular growth and in repeated near- and far- work. And it is not necessary to invoke more

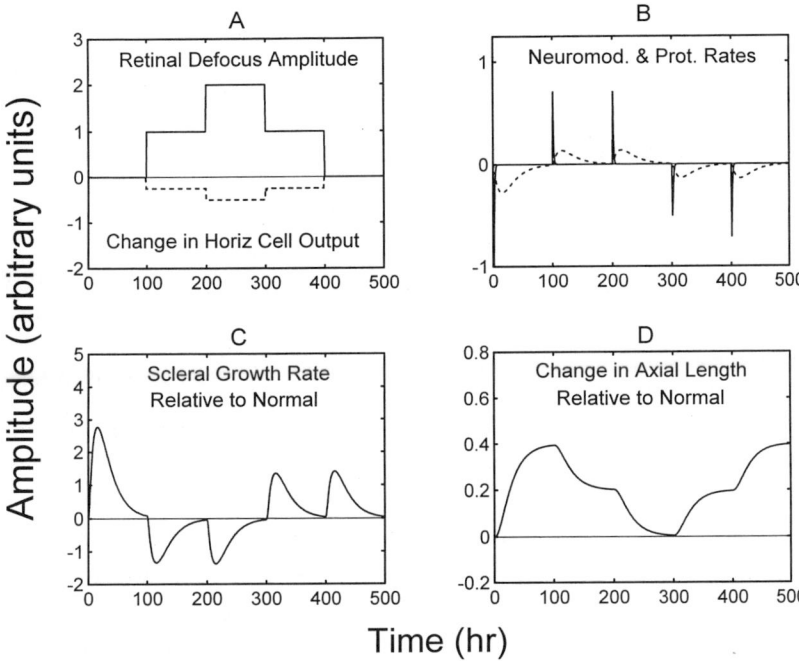

Fig. 34 (A) Envelope of surround stimulus representing various levels of defocus (solid). Changes in horizontal cell output, which is regulated by interplexiform neuronal feedback, shows a complementary response to surround defocus (dashed). (B) Pulses of rates of neuromodulator release (solid) and proteoglycan synthesis (dash) occur at the transitions of surround stimulus (see A). The initial negative pulse (dashed) is due to model parameter start-up transient response to a small retinal-defocus magnitude. (C) Rate of scleral growth relative to normal follows the pulses in the rate of proteoglycan synthesis. (D) Integration of scleral growth rate provide the change in axial length relative to normal. The direction of change is consistant with experiments and with the analysis provided by the schematic model (reprinted from Hung and Ciuffreda,[69] with permission).

complicated processes, such as sensing and analyzing of chromatic aberration, spherical aberration, spatial gradient of blur or spatial frequency content.[15,16] Thus, the Incremental Retinal-Defocus Theory provides increased understanding of the underlying retinal mechanisms for detecting blur magnitude. Furthermore, it explains how this signal is processed to modulate the rate of eye growth, and in turn the resultant development of axial myopia. Finally, a MATLAB/SIMULNK model of the retinal pathway for regulating ocular growth demonstrates quantitatively how change in retinal-defocus magnitude can result in appropriate changes in axial growth.

Saccade-Vergence Interactions Dynamic Model

The experiments and models discussed so far were concerned with binocular oculomotor responses to symmetric target displacements in depth. Thus, only vergence and accommodative responses were analyzed. However, if the target displacements were asymmetrical, as often occurs in daily life, both saccade and vergence must be analyzed. Accommodation has a much longer latency (~350 msec) than those for saccades (~200 msec) and vergence (~180 msec), and therefore plays a relatively minor role in these early-occurring dynamic interactions. Saccadic and vergence eye movements direct gaze laterally and in depth, respectively. Indeed, the interactions between saccade and vergence have been of considerable interest in recent years because of the hypothesis by some investigators that saccades facilitated vergence dynamics, and the extent of the facilitation was based on the visual scene.[26,37,38,172]

It has been found that under both free-space (FS)[27] and instrument-space (IS)[76,83,174] viewing environments, oculomotor response to a pure conjugate target displacement exhibited a transient divergence during its dynamic time course.[27] These transient divergent movements were seen as large loops in the binocular fixation top-view plot, which deviated from the classical iso-vergence curves.[119] On the other hand, the response to a pure disjunctive target displacement showed essentially a slow symmetric saccade-free vergence movement.[76] Moreover, responses to asymmetrical target displacements also exhibited transient divergence, mainly for far-to-near target displacement and with the occurrence of large divergence transients being greater under the FS[37,38,172] than the IS.[71] Some investigators have proposed that central neural mechanisms are involved in these facilitated velocity disjunctive transients.[37,38,172] For these asymmetrical target responses, a movement was considered facilitated if its peak velocity was higher than that of a pure vergence step response of the same amplitude.[71,83]

To analyze saccade-vergence interactions in detail, experiments were conducted under both FS[26] and IS[174] viewing environments. The FS environment corresponds to the natural viewing of objects in a scene that consists of all the usual cues such as blur, disparity, size, perspective and overlap, whereas the IS environment corresponds to a more restricted viewing of targets in an optical assembly that consists only of disparity cues to the two eyes.[76] Dynamic trajectories and latency differences of responses were obtained for randomly presented asymmetrical targets under both FS and IS environments. Also, a dynamic MATLAB/SIMULINK model was developed to account for the experimental results.

Five visually normal adult subjects, ranging in age from 26 to 49 years, participated in the experiments. Horizontal movements of both eyes were measured using a Skalar infrared eye movement monitor (Model 6500). This eye movement system has a linear range of ± 25 deg, a resolution of 5 min of arc and a bandwidth of 200 Hz. Both stimulus generation and response recording were performed using a PC. For each trial, the response over a 2.5 sec interval following the start of the stimulus was digitized by the online computer at a sampling frequency of 200 Hz. The results for representative data under the FS and IS environments were plotted as time traces as well as top-view trajectories.

A dynamic MATLAB4.2 / SIMULINK 1.3 model based on an earlier schematic model of saccade-vergence interactions was developed[67] (Fig. 35). It consisted of a conjugate pulse-step controller and a step disjunctive controller. The model was used to simulate eye movement responses to −4 and −8 deg target displacements in the left and right eyes respectively, corresponding to −6 deg of conjugate ((LE+RE)/2) and 4 deg of disjunctive (LE−RE) stimuli. The latency difference between conjugate and disjunctive components was set at either 0 or 100 msec, with the disjunctive latency held at 200 msec. The eye movements were graphed as time traces and top-view plots for comparison with experimental results. In addition, to examine the effect of model parameter variations, simulations were performed for: pure 10, 20 and 30 deg rightward conjugate target displacements; asymmetrical target displacements requiring a 20 deg rightward conjugate and 10 deg convergence movement; and asymmetrical target displacements requiring a 20 deg leftward conjugate and 10 deg divergence movement. For the latter two conditions, latency for vergence was held at 200 msec, while that for the saccade was varied at 200, 250 and 300 msec. For all these simulations, the binocular fixation trajectories were graphed as top-view plots.

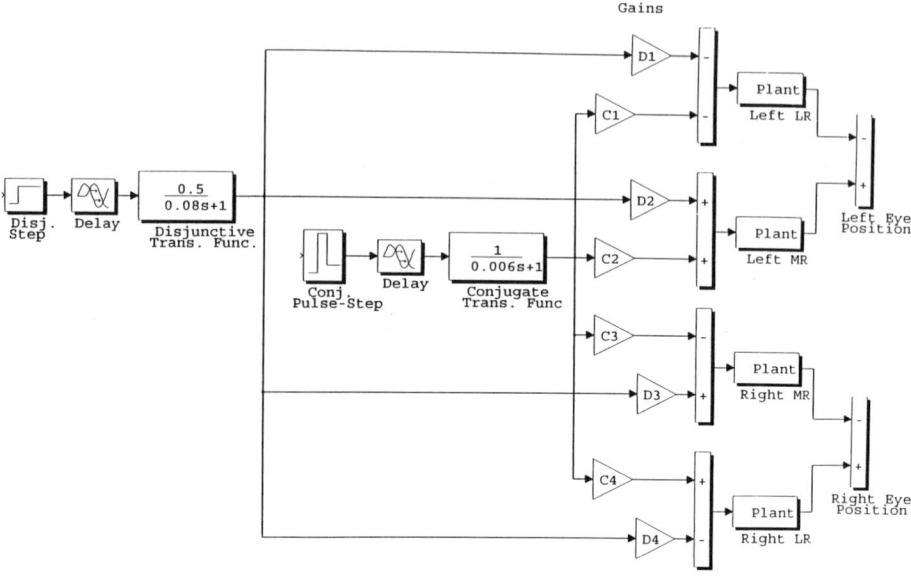

Fig. 35 Dynamic saccade-vergence model showing conjugate pulse-step and disjunctive step controllers. Conjugate and disjunctive transfer functions are used to provide appropriate overall dynamics for saccade and vergence, respectively. Each conjugate (C1 to C4) and disjunctive (D1 to D4) gain has a nominal value of 0.5. The extraocular muscle plant is based on that by Robinson[129] and Zuber and Stark,[175] and is given by $1/(0.064s+1)$ $(0.007s+1)$. An important characteristic of the model is that the conjugate signal to the MR of the contralateral adducting eye is 6 msec longer than that for the LR of the ipsilateral abducting eye. For example, for a rightward saccade, the delay at C2 is 6 msec longer than that at C4. This delay is crucial in producing the saccadic-induced transients in the disjunctive velocity trace (see Fig. 36). A first-order filter with a time constant of 20 msec is placed following the delay stage to provide smooth transients. The model simulations are consistent with physiological results[103] (reprinted from Hung,[67] p. 10, with permission).

The experimental results showed that the average latency difference between saccade and vergence for five visually normal subjects under the FS environment was -10.5 ± 14.8 msec (t = -1.6, d.f. = 4, P > 0.05), and was therefore not significantly different from zero. This indicated that the saccade and vergence controller signals occurred simultaneously for random target presentations under the FS environment. On the other hand, under the IS environment, the average latency difference was 35.9 ± 15.7 msec (t = 5.1,

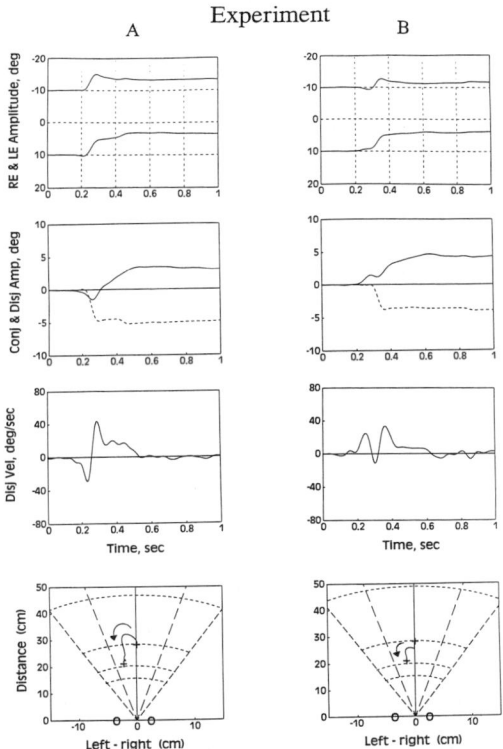

Fig. 36 (**A**) Representative experimental time traces under the free-space (FS) environment for stimulus requiring a response of −4 deg in the LE and −8 deg in the RE, corresponding to 4 deg of convergence and 6 deg of leftward versional movement (Subj. SW). (**B**) Representative experimental time traces under the instrument space (IS) environment for stimulus requiring a response of −2 deg in the LE and −6 deg in the RE, corresponding to 4 deg of convergence and 4 deg of leftward versional movement (Subj. GH). Top graph — left eye (LE, upper) and right eye (RE, lower) time traces. Positive number represents rightward movement. Second graph — conjugate (dotted) and disjunctive (solid) amplitude time courses. Third graph — disjunctive velocity time course. Bottom graph — top-view binocular fixation trajectories corresponding to the movements shown in top graph. The initial central fixation point and the target are shown as "+" symbols. The circular-shaped iso-vergence arcs (dotted) are separated at 5 deg intervals, whereas the radial lines (dashed) are separated at 10 deg intervals. Note that for the bottom graph under the FS environment (A), the trajectory, starting from a position indicated by the central fixation cross, consists of an overshoot loop followed by a radially-directed vergence movement towards the target. On the other hand, under the IS environment (B), the trajectory consists of an initial convergence (along the central radial line), followed by a saccadic trajectory, which is then followed by a final convergence movement (along another radial line) (reprinted from Hung,[67] p. 12, with permission).

d.f. = 4, P < 0.01), which was significantly different from zero. This indicated that vergence generally occurred before the saccade for random target presentation under the IS environment.

Figures 36A and B show representative time traces for experimental data under the (A) FS and (B) IS environments. Note that in the disjunctive velocity trace under the FS environment (Fig. 36, third graph), the saccade-related velocity spike obscured the underlying and ongoing slow smooth vergence velocity peak. On the other hand, under the IS environment (Fig. 36B, third graph), the saccade-related velocity spike occurred later and thereby the envelope (i.e. not including the transient) of the velocity trace revealed more readily the underlying vergence velocity peak.

The corresponding top-view trajectories are shown in bottom graphs of Figs. 36A and B. The top-view trajectory has recently become a standard means for displaying the trace of the binocular fixation point (i.e. the intersection between the lines of sight of the two eyes in space) over the eye movement time course, with the circular-shaped iso-vergence arcs (dotted) corresponding to pure versional eye movements and the radial lines (dashed) corresponding to pure vergence eye movements.[27] It can be seen that under the FS environment (Fig. 36A), the trajectory consists of an overshoot loop followed by a radially-directed vergence movement towards the target. On the other hand, under the IS environment (Fig. 36B), there is an initial vergence movement along the radial line, followed by a saccadic trajectory, which is then followed by a final vergence movement towards the target.

The model simulation responses to target displacement of −2 to the left eye and −6 to the right eye (corresponding to 4 deg disjunctive and −4 deg conjugate stimuli) are shown for simultaneous (Fig. 37A) and sequential (Fig. 37B; latency difference between conjugate and disjunctive controllers = 100 msec) controller onset. Note that the model simulation results under the simultaneous onset condition (Fig. 37A) is consistent with that for experimental trials under the FS environment (Fig. 36A). Also, the model simulation results under the sequential onset condition (Fig. 37B) is consistent with that for experimental trials under the IS environment (Fig. 36B). Thus, similar to that noted in the experimental results, in the model disjunctive velocity trace under the simultaneous onset condition (Fig. 37A, third graph), the saccadic-induced spikes obscured the slow smooth vergence velocity peak. On the other hand, under the sequential onset condition (Fig. 37B, third graph), the spikes occurred later and thereby the envelope (i.e. not including the transient) of the velocity trace revealed the underlying vergence velocity peak.

Overall, four types of experimental top-view binocular trajectories were found. They consisted of: straight, overshoot, undershoot and saccade-vergence.[63] Overshoot and undershoot curves were determined by comparing the traces with distinct version (i.e. along an iso-vergence curve) and vergence movements required to arrive at the target. While the overshoot curves occurred

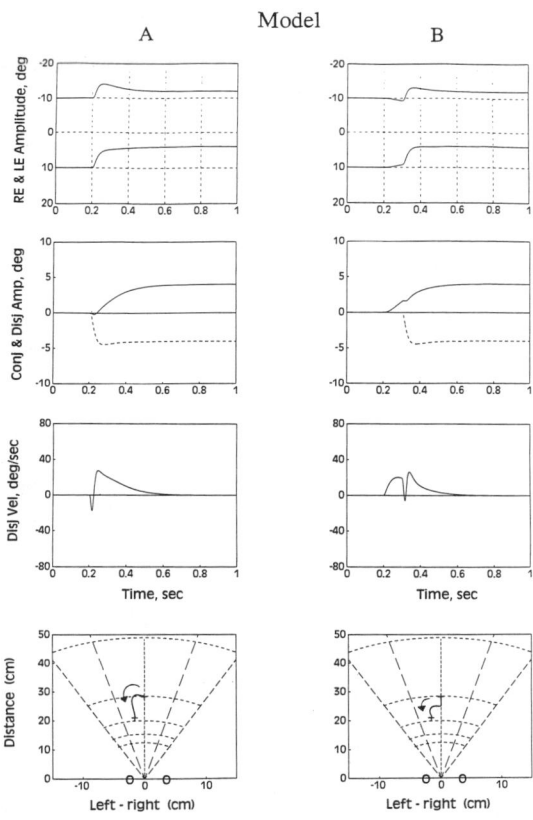

Fig. 37 Model simulation responses for a target displacement requiring −2 deg in the left eye and −6 deg in the left eye, corresponding to 4 deg of convergence and 4 deg of leftward saccadic response for the conditions of **(A)** simultaneous (latency = 200 msec) and **(B)** sequential (latency: disjunctive = 200 msec, and conjugate = 300 msec) onset of controller signals. The description of the traces are the same as those for Fig. 36 (reprinted from Hung,[67] p. 13, with permission).

almost always for near targets, the other curve shapes were seen for both near and far targets.

The percentage of responses for the different trajectory shapes are shown in Fig. 38 for the instrument- and free-space environments in the upper and lower bars, respectively. For each set of bars, the left (light) and right (dark) bars represent near and far targets, respectively. Note the shift towards a predominance of saccade-vergence trajectories, and a scarcity of overshoot trajectories, under the instrument-space as compared to the free-space environment.

Two important timing-related properties were essential in producing the dynamic eye movement characteristics seen in the experimental and model simulation results. First, the small latency difference between signals to the ipsilateral LR and contralateral MR was crucial in producing the transient

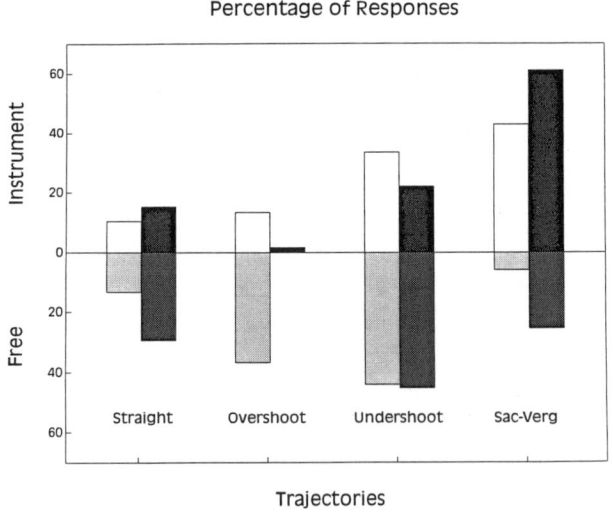

Fig. 38 The percentage of responses of all the subjects combined for the different trajectory shapes (straight, overshoot, undershoot and saccade-vergence) under the instrument (top bars, total $n = 453$) and free-space (bottom bars, total $n = 671$) environments. For each set of bars, the left (light) and right (dark) bars represent near and far targets, respectively. Note the predominance of saccade-vergence movements and a scarcity of overshoot movements under the instrument space, as compared to the free-space environment (reprinted from Hung,[63] p. 163, with permission).

divergence during pure saccades and the saccadic-induced transients in the disjunctive velocity trace (which was reflected in the loops seen in the top-view trajectories) during asymmetrical eye movements. The value of this latency difference was obtained via control simulations that matched the loop shapes in the top-view plots of the experimental data[27] (not shown). The conjugate signal to the MR of the contralateral adducting eye was found to be 6 msec longer than that for the LR of the ipsilateral abducting eye. Thus, for a rightward saccade, the delay at C2 was 6 msec longer than that at C4 (see Fig. 35). This was consistent with the physiological finding of an additional crossover neural connection to the MR of the contralateral eye[103] as well as the experimental finding of differential delays between the two eyes by Smith et al.,[155] who noted that "... small time differences of 0–6 msecs occurred in a consistent way between the two eyes.... The mean value for the right target direction indicated a left eye lag, while that for the left target direction indicated a right eye lag."

Second, the simultaneous onset of the disjunctive and conjugate controller signals could account for the facilitation reported under the FS viewing environment, whereas the sequential onset of the controller signals could account for the lack of facilitation found under the IS viewing environment. This could be seen in both the experimental (Figs. 36A and B) and model simulation (Figs. 37A and B) velocity traces. In the model simulations, both of the velocity traces showed transients during the saccadic interval. However, the key difference was that while the peak of the slow vergence portion of the velocity trace was obscured by the saccadic-velocity transient under the simultaneous condition (Fig. 37A, third graph), it was more readily seen under the sequential condition (Fig. 37B, third graph). This was because in the sequential condition, both the rising and the falling phases of the slow vergence portion of the velocity trace could be seen and its associated velocity peak could be readily discerned. This might explain why under the FS environment (Fig. 36A), corresponding to the simultaneous condition, peak transients could be inadvertently attributed as the peak vergence velocity, resulting in a reporting of enhanced vergence velocities.[37,38,172] On the other hand, under the IS environment (Fig. 36B), corresponding to the sequential condition, peak transients could be readily seen as a saccadic contribution, and the appropriate peak vergence velocity could be obtained. This would result in a reporting of a lack of enhanced vergence velocities.[174]

Also, it should be pointed out that the transient divergence generally did not assist the asymmetrical movements. This can be seen, for example, in the bottom plots of Fig. 37, where the transient divergence moved away from the intended near target. Thus, the transient divergence is in the opposite direction as the required vergence movement.

The consistency of the model simulations and the experimental results in this study indicated that separate conjugate and disjunctive controllers could account for the overall behavior of the responses to asymmetrical targets.[27,71] The key characteristic of the model was that transient divergence during the saccade was accounted for by a delay based on the well-known increase in the contralateral peripheral neural connection.[103] Also, the simultaneous onset of disjunctive and conjugate controllers could account for the greater reporting of enhanced vergence velocities under the FS environment, and the sequential onset of disjunctive and conjugate controllers could account for the lesser reporting enhanced vergence velocities under the IS environment. Indeed, no central computation was needed in the model to determine the amount of "facilitation" required for each change in asymmetrical target position.[37,38,172] Moreover, if such computations were to take place repeatedly in daily life, it would have been an exhausting physiological process. Thus, the parsimony and simplicity offered by the present model reflected the essence of Hering's law,[56] which stated that the two eyes acted as one, so that the separate conjugate and disjunctive controllers worked together to drive the eyes toward the target in space.

Summary Remarks

This monograph has provided an extensive overview of the application of bioengineering techniques to the study of oculomotor control. In the accommodation and vergence systems, static linear and nonlinear elements serve important roles in shaping the steady-state responses and provide insight into clinical abnormalities. For example, decreased lens saturation level while maintaining normal ACG accurately simulates the AS/AR curve as a function of age, and supports the Hess-Gullstrand theory of presbyopia. Also, decreased accommodative controller gain and increased DOF are associated with the visual deficits of amblyopia and congenital nystagmus, respectively. Moreover, high accommodative-convergence or convergence-accommodation crosslink gain can lead to eye alignment deviations, called strabismus or crossed-eye.

Dynamic characteristics of these models have provided important insights into how these systems attain both stability and rapid motor responsivity. When each system is studied in isolation, its response characteristics provide fundamental clues regarding the system's neural control strategy. For both the accommodation and vergence systems, whose latencies are long relative to their dynamics, a continuous feedback control process would lead to instability oscillations. It turns out that the strategy used is to respond with an initial fast open-loop movement that provides a large portion of the response amplitude, followed by a slow closed-loop movement that reduces the residual error to a minimum. In this way, dynamic responsivity and accuracy is attained without introducing instability oscillations. A similar strategy is used by the saccadic system. The saccadic movement is driven by open-loop control, which is followed by secondary saccades to reduce the residual error. Thus, very rapid dynamics is achieved while maintaining accuracy and stability.

However, when these systems operate together, as is generally the case in daily life, their responses are not just simple summations of their isolated motor responses. For example, the neural linkage between the accommodation and vergence control processes results in a combined dual-interactive feedback control system that is quite complex. Additional complexity is introduced when the model is used to investigate proximal and adaptive control of accommodation and vergence. It was shown that these problems could be solved using engineering feedback control systems techniques.

The process of emmetropization had posed a dilemma for previous proposed mechanisms due to the even-error nature of retinal-defocus. Without a sign difference in the blur signal, the retina should not be able to determine the direction for axial growth. However, the Incremental Retinal-Defocus Theory resolves this dilemma by proposing that the change in blur magnitude during an increment of genetically programmed axial growth provides the necessary sign for growth. Retinal circuitry respond to this change in blur magnitude by either increasing or decreasing the rate of proteoglycan synthesis, which in turn modulate appropriately the rate of scleral growth, resulting in emmetropization. Moreover, the theory can be applied to prolonged nearwork, where it can be shown that the resulting decrease in the rate of proteoglycan synthesis causes an increase in axial growth rate, which eventually leads to the development of myopia.

In the saccadic and vergence systems, the finding of dynamic interactions had led to some confusion regarding the underlying control processes. The primary problem is that both saccade and vergence share the same motor output, namely the extraocular muscles that rotate the eyeballs. Hence, their individual contributions must be inferred from the eye movement responses. It had been suggested in recent years by some investigators that vergence responses are facilitated by saccades depending on the characteristics of the visual scene. It turns out that the transient vergence contribution during a saccade is due to a small difference between peripheral neural delays of the signals to the two eyes, and is not due to differences in the visual scene. Indeed, the resolution of this problem involved experimentation using specially designed paradigms and analysis techniques, and simulations using a uniquely configured model. The results support Hering's law, which states that the two eyes acted as one, so that the separate conjugate and disjunctive controllers worked together to drive the eyes toward the target in space.

In the past two to three decades, much has been learned from these models regarding oculomotor control processes. Future extensions of the models include the detailed quantitative investigations of the development of myopia and ocular abnormalities such as strabismus and amblyopia, the effect of training using auditory biofeedback on the normalization of oculomotor function, and the interactions between visual, auditory and perceptual cues in complex multimedia displays.

References

1. Alpern M. Vergence and Accommodation: I, Can change in size induce vergence eye movements? *Arch. Ophthal.* 1958; 60: 355–357.
2. Alpern M. Types of movements. In *The Eye, Muscular Mechanisms* (Academic Press, New York, 1969), Vol. 3. Chap. 5, pp. 5–12 and 141–156.
3. Bahill AT, Stark L. The trajectories of saccadic eye movements. *Sci. Am.* 1979; 240: 108–117.
4. Birnbaum MH. *Optometric Management of Nearpoint Vision Disorders* (Butterworth-Heinemann, Boston, 1993), pp. 303–309.
5. Borish IM. *Clinical Refraction*, 3rd. ed. (Professional Press, Chicago, IL, 1970).
6. Bowman DK, Kertesz AE. Fusional responses of strabismics to foveal and extrafoveal stimulation. *Invest. Ophthal. Vis. Sci.* 1985; 26: 1731–1739.
7. Campbell FW. The accommodation response of the human eye. *Br. J. Physiol. Opt.* 1959; 16: 188–203.
8. Campbell FW. Correlation of accommodation between the two eyes. *J. Opt. Soc. Am.* 1960; 50: 738.
9. Campbell FW. The depth of field of the human eye. *Opt. Acta* 1957; 4: 157–164.
10. Campbell FW, Robson JG, Westheimer G. Fluctuations of accommodation under steady viewing conditions. *J. Physiol.* 1959; 145: 579–594.
11. Campbell FW, Westheimer G. Dynamics of accommodation responses of the human eye. *J. Physiol.* 1960; 151: 285–295.
12. Charman WN, Tucker J. Accommodation as a function of object form. *Am. J. Optom. Physiol. Opt.* 1978; 55: 84–92.
13. Chauhan K, Charman WN. Single figure indices for the steady-state accommodative response. *Ophthal. Physiol. Opt.* 1995; 15: 217–221.
14. Christensen AM, Wallman J. Evidence that increased scleral growth underlies visual deprivation myopia in chicks. *Invest. Ophthal. Vis. Sci.* 1991; 32: 2143–2150.
15. Ciuffreda KJ. Accommodation, pupil, and presbyopia. In *Borish's Clinical Refraction*, ed. W. J. Benjamin (W. B. Saunders Co., Philadelphia, PA, 1998), Chap. 4, pp. 77–120.

16. Ciuffreda KJ. Accommodation and its anomalies. In *Visual Optics and Instrumentation,* Vol. 1, ed. W. N. Charman (Macmillan, London, 1991), pp. 231–279.
17. Ciuffreda KJ, Hokoda SC, Hung G, Semmlow JL. Accommodative stimulus/response function in human amblyopia. *Docum. Ophthal.* 1984; 56: 303–326.
18. Ciufffreda KJ, Hung G. Symptoms related to abnormal tonic state: experimental results and computer simulations. *Optom. Vis. Sci.* 1992; 69: 283–288.
19. Ciuffreda KJ, Kenyon RV. Accommodative vergence and accommodation in normals, ambplyopes, and strabismics. In *Vergence Eye Movements: Basic and Clinical Aspects,* eds. C. M. Schor and K. J. Ciuffreda (Butterworths, Boston, MA. 1983), Chap. 5, pp. 101–173.
20. Ciuffreda KJ, Kellndorfer J, Rumpf D. Contrast and accommodation. In *Presbyopia, Recent Research and Reviews from the 3rd International Symposium* (Professional Press, New York, 1987), Chap. 15, pp. 116–122.
21. Ciuffreda KJ, Levi DM, Selenow A. *Amblyopia: Basic and Clinical Aspects* (Butterworths, Boston, 1992).
22. Ciuffreda KJ, Rosenfield M, Chen H-W. "The AC/A ratio, age and presbyopia," *Ophthal. Physiol. Opt.* 1997; 17: 307–315.
23. Ciuffreda KJ, Rosenfield M, Mordi J, Chen H-W. "Accommodation, age and presbyopia," in *Accommodative and Vergence Interactions,* ed. O. Fransen (Springer-Verlag, New York, 2000), pp. 193–200.
24. Ciuffreda KJ, Tannen B. *Eye Movement Basics for the Clinician* (Mosby, St. Louis, 1995), Chap. 8, pp. 184–205.
25. Ciuffreda KJ, Wallis D. Myopes exhibit increased susceptibility to nearwork-induced transient myopia. *Invest. Ophthal. Vis. Sci.* 1998; 39: 1797–1803.
26. Collewijn H, Erkelens CJ, Steinman RM. Voluntary binocular gaze-shifts in the plane of regard: dynamics of version and vergence. *Vis. Res.* 1995; 35: 3335–3358.
27. Collewijn H, Erkelens CJ, Steinman RM. Trajectories of the human binocular fixation point during conjugate and non-conjugate gaze shifts. *Vis. Res.* 1997; 37: 1049–1069.
28. Cornell E. The influence of orthoptic treatment on proximal convergence. *Aust. Orthop. J.* 1979; 17: 30–32.
29. Cornsweet TN, Crane HD. Servo-controlled infrared optometer. *J. Opt. Soc. Am.* 1970; 60: 548–554.
30. Curtin BJ, *The Myopias: Basic Science and Clinical Management* (Harper and Row, Philadelphia, PA, 1985), pp. 61–151.
31. D'Azzo JD, Houpis CH. *Feedback Control Systems Analysis and Synthesis* (McGraw-Hill, New York, 1988).

32. Duane A. The accommodation and Donder's curve and the need of revising our ideas regarding them. *J. Am. Med. Assoc.* 1909; 52: 1992–1996.
33. Duane A. Normal values of the accommodation at all ages. *Trans. Ophthal. Amer. Med. Assoc.* 1912: 383–391.
34. Duckman RH. The incidence of visual anomalies in a population of cerebral palsied children. *J. Am. Optom. Assoc.* 1979; 50: 1013–1016.
35. Duhamel JR, Colb CR, Goldberg ME. The updating of the representation of visual space in parietal cortex by intended eye movements. *Science* 1992; 255: 90–92.
36. Ehrlich DL. Near vision stress: vergence adaptation and accommodative fatigue. *Ophthal. Physiol. Opt* 1987; 7: 353–357.
37. Enright JT. Changes in vergence mediated by saccades. *J. Physiol.* 1984; 350: 9–31.
38. Erkelens CJ, Van der Steen J, Steinman RM, Collewijn H. Ocular vergence under natural conditions: II. Gaze shifts between targets differing in distance and direction. *Proc. Royal Soc. London (B)* 1989; 236: 441–465.
39. Fincham EF. The accommodative reflex and its stimulus. *Br. J. Ophthal.* 1951; 35: 381–393.
40. Fincham EF, Walton J. The reciprocal actions of accommodation and vergence. *J. Physiol.* 1957; 137: 488–508.
41. Fisher RF. The force of contraction of the human ciliary muscle during accommodation. *J. Physiol.* 1977; 270: 51–74.
42. Fisher SK, Ciuffreda KJ. Accommodation and apparent distance. *Perception* 1988; 17: 609–621.
43. Fisher SK, Ciuffreda KJ, Levine S. Tonic accommodation, accommodative hysteresis and refractive error. *Am. J. Optom. Physiol. Opt.* 1987; 64: 799–809.
44. Flitcroft DI. A model of the contribution of oculomotor and optical factors to emmetropization and myopia. *Vis. Res.* 1998; 38: 2869–2879.
45. Fry GA. Blur of the retinal image. *Br. J. Physiol. Opt.* 1955; 12: 130–152.
46. Fry GA. *Handbook of Physiology — Neurophysiology I* (American Physiological Society, 1959), Chap. 27, pp. 647–670.
47. Fulk GW, Cyert LA, Parker DE. Baseline characterstics in the myopia progression study, a clinical trial of bifocals do slow myopia progression. *Optom. Vis. Sci.* 1998; 75: 485–492.
48. Goss DA, Wickham MG. Retinal-image mediated ocular growth as a mechanism for juvenile onset myopia and for emmetropization. *Docum. Ophthal.* 1995; 90: 341–375.

49. Griffin JR. *Binocular Anomalies — Procedures for Vision Therapy* (Professional Press, Chicago IL, USA, 1976).
50. Grosvenor T, Flom MC (eds.). *Refractive Anomalies — Research and Clinical Applications* (Butterworth-Heinemann, Boston, 1991).
51. Gwiazda J, Thorn F, Bauer J, Held R. Myopic children show insufficient accommodative response to blur. *Invest. Ophthal. Vis. Sci.* 1993; 34: 690–694.
52. Hamasaki D, Ong J, Marg E. The amplitude of accommodation. *Am. J. Optom. Arch. Am. Acad. Optom.* 1956; 33: 3–14.
53. Heath G. The influence of visual acuity on accommodative responses of the eye. *Am. J. Optom. Arch. Am. Acad. Optom.* 1956; 33: 513–524.
54. Henson D. *Optometric Instrumentation* (Butterworths, Woburn, MA. 1983), Chap. 8, pp. 152–162.
55. Henson DB, North R. Adaptation to prism-induced herophoria. *Am. J. Optom. Physiol. Opt.* 1980; 57: 129–137.
56. Hering E. In *The Theory of Binocular Vision*, eds. B. Bridgman and L. Stark trans. from the 1868 German original (Plenum Press, New York, 1977), pp. 50–55.
57. Hess C. Beobachtungen ueber den Akkommodationsvorgang. *Klin. Mbl. Augenheilk.* 1904; 42: 309–315.
58. Hofstetter HW. The relationship of proximal convergence to fusional and accommodative convergence. *Am. J. Optom. Arch. Am. Acad. Optom.* 195; 28: 300–308.
59. Hofstetter HW. A longitudinal study of amplitude of changes in presbyopia. *Am. J. Optom. Arch. Am. Acad. Optom.* 1965; 42: 3–8.
60. Hokoda SC, Ciuffreda KJ. Theoretical and clinical importance of proximal vergence and accommodation. In *Vergence Eye Movements: Basic and Clinical Aspects*, eds. C. M. Schor and K. J. Ciuffreda (Butterworths, Boston, MA, 1983), pp. 75–97.
61. Hugonnier R, Hugonnier SC. *Strabismus, Heterophoria, Ocular Motor Paralysis*, trans. Troutman SV (C. V. Mosby, St. Louis, MO, 1969).
62. Hung G. Dynamic model of the vergence eye movement system: simulation using MATLAB/SIMULINK. *Comp. Meth. Prog. Biomed.* 1998; 55: 59–68.
63. Hung G. Saccade-vergence trajectories under free- and instrument-space environments. *Curr. Eye Res.* 1998; 17: 159–164.
64. Hung G. Quantitative analysis of associated and disassociated phorias: linear and nonlinear static models. *IEEE Trans. Biomed. Eng.* 1992; 39: 135–145.
65. Hung G. Quantitative analysis of the accommodative convergence to accommodation ratio: linear and nonlinear static models. *IEEE Trans. Biomed. Eng.* 1997; 44: 306–316.

66. Hung G. Sensitivity analysis of the stimulus-response function of a static nonlinear accommodation model. *IEEE Trans. Biomed. Eng.* 1998; 45: 335–341.
67. Hung G. Dynamic model of saccade-vergence interactions. *Med. Sci. Res.* 1998; 26: 9–14.
68. Hung G. Adaptation model of accommodation and vergence. *Ophthal. Physiol. Opt.* 1992; 12: 319–326.
69. Hung G, Ciuffreda KJ. A unifying theory of refractive error development. *Bull. Math. Biol.* 2000; 62: 1087–1108.
70. Hung G, Ciuffreda KJ. Sensitivity analysis of relative accommodation and vergence. *IEEE Trans. Biomed. Eng.* 1994; 4: 241–248.
71. Hung G, Ciuffreda KJ. Schematic model of saccade-vergence interactions. *Med. Sci. Res.* 1996; 24: 813–816.
72. Hung G, Ciuffreda KJ. Dual-mode behavior in the human accommodation system. *Ophthal. Physiol. Opt.* 1988; 8: 327–332.
73. Hung G, Ciuffreda KJ. Adaptation model of nearwork-induced transient myopia. *Ophthal. Physiol. Opt.* 1999; 19: 151–158.
74. Hung G, Ciuffreda KJ, Rosenfield M. Proximal contribution to a linear static model of accommodation and vergence. *Ophthal. Physiol. Opt.* 1996; 16: 34–41.
75. Hung G, Ciuffreda KJ, Semmlow JL. Static vergence and accommodation: population norms and orthoptics effects. *Docum. Ophthal.* 1986; 62: 165–179.
76. Hung G, Ciuffreda KJ, Semmlow JL, Horng J-L. Vergence eye movements under natural viewing conditions. *Invest. Ophthal. Vis. Sci.* 1994; 35: 3486–3492.
77. Hung G, Semmlow JL. Static behavior of accommodation and vergence: computer simulation of an interactive dual-feedback system. *IEEE Trans. Biomed. Eng.* 1980; 27: 439–447.
78. Hung G. Semmlow JL. A quantitative theory of control sharing between accommodative and vergence controllers. *IEEE Trans. Biomed. Eng.* 1982; 29: 364–370.
79. Hung G, Semmlow JL, Ciuffreda KJ. The near response: modeling, instrumentation, and clinical applications. *IEEE Trans. Biomed. Eng.* 1984; 31: 910–919.
80. Hung G, Semmlow JL, Ciuffreda KJ. A dual-mode dynamic model of the vergence eye movement system. *IEEE Trans. Biomed. Eng.* 1986; 33: 1021–1028.
81. Hung G, Semmlow JL, Ciuffreda KJ. Accommodative oscillation can enhance average accommodative response: a simulation study. *IEEE Trans. Syst. Man Cybern.* 1982; 12: 594–598.

82. Hung G, Semmlow JL, Ciuffreda KJ. Identification of accommodative vergence contribution to the near response using response variance. *Invest. Ophthal. Vis. Sci.* 1983; 24: 772–777.
83. Hung G, Zhu H-M, Ciuffreda KJ. Convergence and divergence exhibit different response characteristics to symmetric stimuli. *Vis. Res.* 1997; 37: 1197–1205.
84. Ittelson WH. *Visual Space Perception* (Springer, New York, 1960), pp. 151–168.
85. Javitt JC, Chiang YP. The socioeconomic aspects of laser refractive surgery. *Arch. Ophthal.* 1994; 112: 1526–1530.
86. Jiang B-C. Integration of a sensory component into the accommodation model reveals differences between emmetropia and late-onset myopia. *Invest. Ophthal. Vis. Sci.* 1997; 38: 1511–1516.
87. Johnson CA. Effects of luminance and stimulus distance on accommodation and visual resolution. *J. Opt. Soc. Am.* 1976; 66: 138–142.
88. Jones R. The effect of proximal accommodation on accommodative accuracy. *Optom. Vision Sci. (Suppl.).* 1993; 70: 56.
89. Kenyon RV, Ciuffreda KJ, Stark L. Dynamic vergence eye movements in strabismus and amblyopia: symmetric vergence. *Invest. Ophthal. Vis. Sci.* 1980; 19: 60–74.
90. Kowler E, Steinman RM. The effect of expectations on slow oculomotor control — I. Periodic target step. *Vis. Res.* 1979; 19: 619–632.
91. Krall AM, Fornaro R. An algorithm for generating root locus diagrams. *Commun. Assoc. Comp. Machin.* 1967; 10: 186–188.
92. Krishnan VV, Farazian F, Stark L. An analysis of latencies and prediction in the fusional vergence system. *Am. J. Optom. Arch. Am. Acad. Optom.* 1973; 50: 933–939.
93. Krishnan VV, Stark L. Integral control in accommodation. *Comp. Prog. Biomed.* 1975; 4: 237–245.
94. Krishnan VV and Stark L. A heuristic model for the human vergence eye movement system. *IEEE Trans. Biomed. Eng.* 1977; 24: 44–49.
95. Krommenhoek KP, van Opstal AJ, Gielen CCAM, van Gisbergen JAM. Remapping of neural activity in the motor colliculus: a neural network study. *Vis. Res.* 1993; 33: 1287–1298.
96. Lancaster WZ, Williams ER. New light on the theory of accommodation with practical applications. *Trans. Am. Acad. Ophthal. Otolaryngol.* 1914; 19: 170–195.
97. Last RJ (ed.). *Wolff's Anatomy of the Eye and Orbit* (W. B. Saunders Co., Philadelphia, PA, 1968), Chap. 2, p. 30.

98. Leigh RJ, Zee DS, *The Neurology of Eye Movements* (Philadelphia, PA, F. A. Davies Co., 1983).
99. Liebowitz HW, Owens DA. Night myopia and the intermediate dark focus of accommodation. *J. Opt. Soc. Am.* 1975; 65: 1121–1128.
100. Lin LLK, Shih YF, Lee YC, Hung PT, Hou PK. Changes in ocular refraction and its components among medical students — a 5-year longitudinal study. *Optom. Vis. Sci.* 1996; 73: 495–498.
101. Maddox E. Investigations in the relation between convergence and accommodation of the eyes. *J. Anat. Physiol.* 1886; 20: 475–508 and 565–584.
102. Mahlman HE. *Handbook of Federal Vision Requirements and Information* (Professional Press, Chicago, IL, 1982).
103. Mays LE. Neurophysiological correlates of vergence eye movements. In *Vergence Eye Movements: Basic and Clinical Aspects*, eds. C. M. Schor and K. J. Ciuffreda (Butterworths, Boston, MA, 1983), Chap. 20, pp. 649–670.
104. McBrien NA, Millidot M. The effect of refractive error on the accommodative response gradient. *Ophthal. Physiol. Opt.* 1986; 6: 145–149.
105. Medina A, Fariza E. Emmetropization as a first-order feedback system. *Vis. Res.* 1993; 33: 21–26.
106. Merchant J, Morrissette R, Porterfield JL. Remote measurement of eye direction allowing subject motion over one cubic of space. *IEEE Trans. Biomed. Eng.* 1974; 21: 309–317.
107. Miller PJ. *Dynamics of Voluntary Vergence in Intermittent Exotropia*, M.S. Thesis (University of California at Berkeley, Berkeley, CA, 1973).
108. Mordi JA, Ciuffreda KJ. Static aspects of accommodation: age and presbyopia. *Vis. Res.* 1998; 38: 1643–1653.
109. Morgan MW. The resting state of accommodation. *Am. J. Optom. Arch. Am. Acad. Optom.* 1957; 34: 347–353.
110. Morgan MW. Accommodation and vergence. *Am. J. Optom. Arch. Am. Acad. Optom.* 1968; 45: 417–454.
111. Moses RA (ed.). *Adler's Physiologiy of the Eye, Clinical Applications* (C. V. Mosby Co., St. Louis, MO, 198), Chap. 5, p. 92.
112. North RV, Henson DB, Smith TJ. Influence of proximal accommodative and disparity stimuli upon the vergence system. *Ophthal. Physiol. Opt.* 1993; 13: 239–243.
113. Norton TT, Rada JA. Reduced extracellular matrix in mammalian sclera with induced myopia. *Vis. Res.* 1995; 35: 1271–1281.
114. Ogle KN, Martens TG, Dyer JA. *Binocular Vision and Fixation Disparity* (Lea and Febiger, Philadelphia, PA, 1967), Chaps. 2–5, pp. 9–119.

115. O'Neill WD, Sanathanan CK, Brodkey JS. A minimum variance, time optimal, control system model of the human lens accommodation. *IEEE Trans. Syst. Sci. Cybern.* 1969; 5: 290–299.
116. Ong E, Ciuffreda KJ, *Accommodation, Nearwork and Myopia* (Optometric Extension Program, Santa Ana, CA, 1997).
117. Ong E, Ciuffreda KJ. Nearwork-induced transient myopia — a critical review. *Docum. Ophthal.* 1995; 91: 57–85.
118. Ong E, Ciuffreda KJ, Tannen B. Static accommodation in congenital nystagmus. *Invest. Ophthal. Vis. Sci.* 1993; 34: 194–204.
119. Ono H. Saccadic eye movements during changes in fixation to stimuli at different distances. *Vis. Res.* 1977; 17: 233–238.
120. Optican LM, Miles FA. Visually induced adaptive changes in primate saccadic oculomotor control signals. *J. Neurophysiol.* 1985; 54: 940–958.
121. Panum EL. *Physiologische Untersuchungen Über das Sehen mit zwei Augen* (Schwersche Buchhandlung, Kiel, Germany, 1858).
122. Phillips S. *Ocular Neurological Control Systems: Accommodation and Near Response Triad*, Ph.D. Dissertation (University of California at Berkeley, Berkeley, CA, 1974).
123. Poggio GF, Fischer B. Binocular interaction and depth sensitivity in striate and prestriate cortex of behaving Rhesus monkey. *J. Neurophysiol.* 1977; 40: 1392–1405.
124. Ramsdale C, Charman WN. A longitudinal study of the changes in the static accommodation response. *Ophthal. Physiol. Opt.* 1989; 9: 255–263.
125. Rashbass C, Westheimer G. Disjunctive eye movements. *J. Physiol.* 196; 159: 339–360.
126. Reading RW. *Binocular Vision* (Butterworths, Boston, MA, 1983), pp. 1–8.
127. Remmel RS. An inexpensive eye movement monitor using the scleral search coil technique. *IEEE Trans. Biomed. Eng.* 1984; 31: 388–390.
128. Ripps H, Chin NB, Siegel IM, Breinin GM. Effect of pupil size on accommodation, convergence, and the AC/A ratio. *Invest. Ophthal. Vis. Sci.* 1962; 1: 127–135.
129. Robinson DA. Models of the saccadic eye movement control system. *Kybernetic* 1973; 14: 71–83.
130. Rosenfield M, Ciuffreda KJ. Effect of surround propinquity on the open-loop accommodative response. *Invest. Ophthal. Vis. Sci.* 1991; 32: 142–147.
131. Rosenfield M, Ciuffreda KJ. Distance heterophoria and tonic vergence. *Optom. Vision Sci.* 1990; 67: 667–669.

132. Rosenfield R, Ciuffreda KJ, Hung GK. Linearity of proximally-induced accommodation and vergence. *Invest. Ophthal. Vis. Sci.* 1991; 32: 2985–2991.
133. Rosenfield M, Ciuffreda KJ, Hung GK, Gilmartin B. Tonic accommodation: a review. I. Basic aspects. *Ophthal. Physiol. Opt.* 1993; 13: 266–284.
134. Rosenfield M, Ciuffreda KJ, Hung GK, Gilmartin B. Tonic accommodation: a review. II. Accommodative adaptation and clinical aspects. *Ophthal. Physiol. Opt.* 1994; 14: 265–277.
135. Rosenfield M, Ciuffreda KJ, Novogrodsky L. Contribution of accommodation and disparity-vergence to transient nearwork-induced myopic shifts. *Ophthal. Physiol. Opt.* 1992; 12: 433–436.
136. Rosenfield M, Ciuffreda KJ, Rosen J. Accommodative response during distance optometric test procedures. *J. Am. Optom. Assoc.* 1992; 63: 614–618.
137. Rosenfield M, Ciuffreda KJ, Ong E, Azimi A. Proximally-induced accommodation and accommodative adaptation. *Invest. Ophthal. Vis. Sci.* 1990; 31: 1162–1167.
138. Rosenfield R, Gilmartin B (eds.). *Myopia and Nearwork* (Butterworth-Heinemann, Oxford, UK, 1998).
139. Rosenfield M, Gilmartin B. Temporal aspects of accommodative adaptation. *Optom. Vision Sci.* 1989; 66: 229–234.
140. Rosenfield M, Gilmartin B. Effect of target proximity on the open-loop accommodative response. *Optom. Vision Sci.* 1990; 67: 74–79.
141. Saladin JJ, Stark L. Presbyopia: new evidence from impedance cyclography supporting the Hess-Gullstrand theory. *Vis. Res.* 1975; 15: 537–541.
142. Scammon RE, Armstrong EL. On the growth of the human eyeball and optic nerve. *J. Comp. Neurol.* 1925; 38: 165–219.
143. Schaeffel F, Howland HC. Mathematical model of emmetropization in the chicken. *J. Opt. Soc. Am. A.* 1988; 5: 2080–2086.
144. Schaeffel F, Troilo D, Wallman J, Howland HC. Developing eyes that lack accommodation grow to compensate for imposed defocus. *Vis. Neurosci.* 1990; 4: 177–183.
145. Schor CM. Adaptive regulation of accommodative vergence and vergence accommodation. *Am. J. Optom. Physiol. Opt.* 1986; 63: 587–609.
146. Schor CM, Alexander J, Cormack L, Stevenson S. Negative feedback control of proximal convergence and accommodation. *Ophthal. Physiol. Opt.* 1992; 12: 307–318.
147. Semmlow JL, Hung GK. The near response: theories of control. In *Vergence Eye Movements: Basic and Clinical Aspects,* eds. C. M. Schor and K. J. Ciuffreda (Butterworths, Boston, MA, 1983), Chap. 6, pp. 175–195.

148. Semmlow JL, Hung GK, Ciuffreda KJ. Quantitative assessment of disparity vergence components. *Invest. Ophthal. Vis. Sci.* 1986; 27: 558–564.
149. Semmlow JL, Wetzel P. Dynamic contributions of binocular vergence components. *J. Opt. Sci. Am.* 1979; 69; 639–645.
150. Sethi B, North R. Vergence adaptive changes with varying magnitudes of prism-induced disparities and fusional amplitudes. *Am. J. Optom. Physiol. Opt.* 1987; 64: 263–268.
151. Sheedy JE. Fixation disparity analysis of oculomotor imbalance. *Am. J. Optom. Physiol. Opt.* 1980; 57; 632–639.
152. Shirachi D, Liu J, Lee M, Jang J, Wang J, Stark L. Accommodation dynamics. I. Range nonlinearity. *Am. J. Optom. Physiol. Opt.* 1978; 55: 631–641.
153. Shum, PJT, Ko, LS, Ng CL, Lin SL. A biometric study of ocular changes during accommodation. *Am. J. Ophthal.* 1993; 115: 76–81.
154. Smith EL, Hung LF, Harwerth RS. Effects of optically-induced blur on the refractive status of young monkeys. *Vis. Res.* 1994; 34: 293–301.
155. Smith KU, Schmidt J, Putz V. Binocular coordination: feedback synchronization of eye movements for space perception. *Am. J. Optom. Arch Am. Acad. Optom.* 1970; 47: 679–689.
156. Sorsby A, Leary GA. A longitudinal study of refraction and its components during growth. *Med. Res. Council Special Report Series No. 309* Her Majesty's Stationery Office, London, 1970.
157. Sperduto RD, Seigel D, Roberts J, Rowland M. Prevalence of myopia in the United States. *Arch. Ophthal.* 1983; 101: 405–407.
158. Stark L. *Neurological Control Systems: Studies in Bioengineering* (Plenum, New York, 1968), pp. 60–62 and 236–270.
159. Stark L. *Presbyopia, Recent Research and Reviews From the 3rd International Symposium* (Professional Press, New York, 1987), Chap. 36, pp. 264–274.
160. Strang NC, Winn B, Gilmartin B. Repeatability of post-task regression of accommodation in emmetropia and late-onset myopia. *Ophthal. Physiol. Opt.* 1994; 14: 88–91.
161. Sun F, Stark W. Switching control of accommodation: experimental and simulation responses to ramp inputs. *IEEE Trans. Syst. Sci. Cybern.* 1990; 37: 73–79.
162. Thompson HE. *The Dynamics of Accommodation in Primates*, Ph.D. Dissertation (University of Illinois Medical Center, Chicago, IL, 1975).
163. Toates FM. Vergence eye movements. *Docum. Ophthal.* 1974; 37: 153–214.
164. Toates FM. A model of accommodation. *Vis. Res.* 1970; 10: 1069–1076.

165. Troilo D, Gottlieb MD, Wallman J. Visual deprivation causes myopia in chicks with optic nerve section. *Curr. Eye Res.* 1987; 6: 993–999.
166. Wallman J. Can myopia be prevented? *Research to Prevent Blindness Science Writers Seminar* 1997: 50–52.
167. Westheimer G, Mitchell AM. Eye movement responses to convergence stimuli. *Arch. Ophthal.* 1956; 55: 848–856.
168. Wick B. Clinical factors in proximal vergence. *Am. J. Optom. Physiol. Opt.* 1985; 62: 1–18.
169. Wilson DC. *Accommodation-Accommodative Vergence Synkinesis in the Human Visual System*, Ph.D. Dissertation (University of California at Berkeley, Berkeley, CA, 1972).
170. Winn B, Pugh JR, Gilmratin, Owens H. Arterial pulse modulates steady-state ocular accommodation. *Curr. Eye Res.* 1990; 9: 971–975.
171. Young LR, Sheena D. Survey of eye movement recording methods. *Behav. Res. Meth. Instrum.* 1975; 7: 397–429.
172. Zee DS, Fitzgibbon EJ, Optican LM. Saccade-vergence interactions in humans. *J. Neurophysiol.* 1992; 68: 1624–1641.
173. Zee DS, Levi L. Neurological aspects of vergence eye movements. *Rev. Neurol.* 1989; 145: 613–620.
174. Zhu H-M. *Investigation of Interaction Between Vergence and Saccadic Eye Movements,* M.S. Thesis (Biomedical Engineering, Rutgers University, NJ), pp. 35–38.
175. Zuber BL, Stark L. Dynamic characteristics of fusional vergence eye movements. *IEEE Trans. Syst. Sci. Cybern.* 1968; 4: 72–79.

Index

abducens nucleus 7
accommodative convergence to accommodation (AC/A) ratio 2, 40, 41, 57, 59
    nonlinear (AC/A$_{nonlinear}$ ratio) 46, 49
    response (AC/A$_{response}$ ratio) 36, 37
    stimulus (AC/A$_{stimulus}$ ratio) 36, 37, 39, 43, 44
accommodative stimulus/response (AS/R) function (curve) 24, 26, 29, 31–35, 60, 93–95
adaptation
    model of accommodation and vergence 5, 27, 42, 76–80, 82
    prism 77
amblyopia 26, 32, 59, 60, 110, 112
autocorrelation 69, 73
autorefractometer 17

base-in (BI) prism 39, 47, 48, 56
base-out (BO) prism 39, 47, 48, 56

convergence accommodation to convergence (CA/C) ratio 2, 36, 57
chromatic aberration 5, 101
ciliary muscle 5, 33, 73, 86
conjugate eye movement 7, 102–104, 106, 109, 110, 112

continuous feedback model (system) 13, 26, 64, 69, 74, 111
control system 1, 10, 11, 14, 41
crossover point 32

deadspace operator 14, 23, 41, 44, 46, 54
disjunctive eye movement 7, 103, 104, 106, 109, 110
dopamine 90, 92
dual-mode model (process) 13, 64
    accommodation 5, 73, 74
    vergence 4
Duane-Fincham theory of presbyopia 33

efference copy 54, 55, 70
emmetropia 3
emmetrope (EMM) 82, 83, 86, 87
emmetropization 3, 5, 36, 87, 88, 92, 94, 97
even-error signal 5
extraocular muscle 7
eye tracker (monitor)
    infrared 19, 20
    Iota Eyetrace Systems (Sweden) 19
    ISCAN 19
    scleral search coil 19
    Skalar 103

facilitation 102, 109, 110
fast component 13
    accommodation 74, 77
    vergence (FAST) 64, 70, 77
feedback control system 1, 10–12
fixation disparity (FD) method 39, 41, 46, 49
free-space environment 15, 56, 102–106, 108–110

Hering's law 110, 112
Hess-Gullstrand theory of presbyopia 33, 110
hyperopia 3
hyperope (HYP) 82, 83, 86, 87

Incremental Retinal-Defocus Theory 87, 89, 93, 96, 101, 111
instability oscillation 62
instrument-space (IS) environment 15, 56, 102–106, 108–110
interactive dual-feedback system 36
interplexiform neuron 90, 96–98

lag of accommodation 5, 15, 26, 27, 37, 38, 61, 85, 87, 94
laplace transform (domain) 10–12
lateral rectus muscle 7
lens viewing condition 41, 45

main sequence 61
medial rectus muscle 7
multiple-step response 64, 72, 74
myopia 3, 26, 86–88, 112
myope
    early-onset (EOM) 82, 83, 86
    late-onset (LOM) 82, 83, 86

nearwork-induced transient myopia (NITM) 5, 81–83, 86, 94, 95
neuromodulator 90–92, 98–100
neurotransmitter 90
nystagmus 3, 26, 59, 110

open-loop movement 14
oculomotor balance 35
oculomotor nucleus 7
optically conjugate 18
optometer
    Hardinger coincidence 15, 16
    infrared dynamic 17

phoria 4, 39
    associated 49
    disassociated 49
    method 40, 41, 46, 49
preprogrammed process 64
presbyopia 33, 110
prism viewing condition 41, 44
proteoglycan synthesis 91, 94, 98, 100, 101, 111
proximal
    accommodation 51, 52
    vergence 51, 52

reciprocal innervation 7
refractive error development model 5, 96
residual error 14
retinoscope principle 17, 18
root locus stability analysis 4, 61–63

saccade-vergence interactions dynamic model 5, 102–104
sampled-data behavior 73
Scheiner principle 15, 16

sensitivity analysis
    accommodation model 28, 29
    accommodation and vergence model 54, 56
Sheedy Disparometer 39
slow component 13
    accommodation 74, 75, 77
    vergence (SLOW) 65, 67, 68, 70, 77

spherical aberration 5
strabismus 4, 26, 111, 112

top-view trajectory 105–107

zero-order hold (ZOH) 65, 68, 69, 71
zonular fiber 5